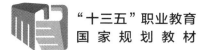

"十三五"职业教育
国家规划教材

高等职业院校
机电类"十三五"规划教材

机电一体化
专业英语
（第3版）

Practical English for
　　Mechatronics (3rd Edition)

黄星 赵诗若 ◎ 主编
陈亚丽 赵九九 王暖 曹杨 ◎ 副主编
赵忠兴 ◎ 主审

北　京

图书在版编目（CIP）数据

机电一体化专业英语 / 黄星，赵诗若主编. -- 3版. -- 北京：人民邮电出版社，2017.7（2022.8重印）
高等职业院校机电类"十三五"规划教材
ISBN 978-7-115-45930-5

Ⅰ．①机… Ⅱ．①黄… ②赵… Ⅲ．①机电一体化－英语－高等职业教育－教材 Ⅳ．①TH-39

中国版本图书馆CIP数据核字(2017)第124747号

内 容 提 要

本书系高职高专专业阶段的英语教材，旨在满足机电一体化及相关专业学生和行业从业人员学习专业英语的需要。

本教材是依照高职高专的培养目标及企业的实际情况，遵循"实用、可用、好用"的原则，在内容编排上突出"贴近企业，贴近实际，贴近岗位"的要求，并力求通俗、易懂、好学、用得上。

全书含10个单元，每个单元都包含一些与生产岗位紧密结合的英语实用实例。

本书既可作为高职高专、成人高校及相关院校的专业英语教材，也可用作企业培训的教材，还可供有关的教师和学生及技术人员学习参考。

◆ 主　编　黄星　赵诗若
　　副主编　陈亚丽　赵九九　王暖　曹杨
　　主　审　赵忠兴
　　责任编辑　李育民
　　责任印制　焦志炜

◆ 人民邮电出版社出版发行　北京市丰台区成寿寺路11号
　　邮编　100164　电子邮件　315@ptpress.com.cn
　　网址　http://www.ptpress.com.cn
　　三河市祥达印刷包装有限公司印刷

◆ 开本：787×1092　1/16
　　印张：16.75　　　　　2017年7月第3版
　　字数：397千字　　　2022年8月河北第13次印刷

定价：44.80元

读者服务热线：(010)81055256　印装质量热线：(010)81055316
反盗版热线：(010)81055315
广告经营许可证：京东市监广登字 20170147 号

(第3版)前言

《机电一体化专业英语(第3版)》是高职高专专业基础阶段的重要学习教材。

依照《高职高专教育英语课程教学基本要求(试行)》的精神,本书继续秉承从培养企业需要的高级应用型人才的总体目标及专业的实际需要出发,结合企业岗位的实际,力求向学生提供更加实用、与岗位衔接更为密切的专业英语教材。

为了更好地满足高职高专院校学生对"机电一体化专业英语"课程学习的需要,本书在充分考虑了专业英语课程兼顾工具性、实用性和科学性特点的同时,更加注重教材的实际应用。本次修订的主要内容如下:

- 对书中个别不妥部分进行了校正与修改,增加了图形练习,更为学生所接受;
- 删除了部分不合理的练习,增加了更为实用的练习;
- 对译文再次进行了完善,力求准确无误。

全书共设 10 个单元,每个单元共设 4 个模块,具体内容见下表。

内　容	要　求
专业阅读 (Technical and Practical Reading)	本部分旨在培养学生专业英语的阅读能力。它包括两篇课文:A 篇精讲,突出机电一体化专业的精髓与特色;B 篇泛学,利用专业课所学知识,借助英语来扩大视野。 所收课文能展示当今机电一体化专业方面的最新发展、最新工艺、最新设备及最新技术等,同时课文附带一些真实场景及实际工作的照片等。每篇课文一般控制在 500~650 字
控制面板概览 (Glance at a Control Panel)	本部分主要帮助学生了解机电一体化(机械与电子)设备的常用知识、部件名称、使用功能及应用范畴,是学生到实训基地与企业实习的必学专业英语内容。 所有内容与企业现场设备及实际应用设备接近,形成衔接,所以这部分内容是本书应用性很强的知识点之一。 "专业认知"是本教材的第一大特色部分
动手能力 (Simulated Writing)	Section A "技能认知"——Match Your Skills,认知企业常用设备及辅助配件。本部分提供了与企业生产紧密相关的资料图片,让学生感受专业英语的应用。Section B "试试您的动手能力"——Having a Try,让学生应用所学常用设备及辅助配件知识进行实际的操作。 同时,还在其他的练习中配有相应的简单产品广告、机床说明书、电子设备操作与使用说明书翻译等应用文。 "动手能力"是本教材的第二大特色部分
扩大视野 (Broaden Your Horizon — Practical Activity)	选用了贴近企业,贴近生产车间,贴近实际的各类简单的工厂实际操作、维修及保养方面的知识进行介绍,可以达到以下效果。第一,可以增强学生学习专业英语的兴趣;第二,学生通过专业英语的学习,可以学到相关专业知识;第三,提升了专业内容与企业实际衔接度;第四,较好地解决了单纯就语言学语言的教学模式。 "扩大视野"是本教材的第三大特色部分

本书由长春汽车工业高等专科学校黄星、赵诗若任主编；漯河职业技术学院的陈亚丽和长春汽车工业高等专科学校赵九九、王暖、曹杨任副主编；赵忠兴任主审。赵忠兴修订了1～6单元译文；赵九九修订了7～10单元译文。王暖编写了第1单元；曹杨编写了第2单元；赵诗若编写了第3单元；宋敏编写了第4单元；高小钠编写了第5单元；陈琳、黄晨路编写了第6单元；孙铭蔚、廖方圆编写了第7单元；陈亚丽编写了第8单元和第9单元；高宪利、穆笑妍编写了第10单元。

本书在修改过程中难免还会存在缺点和错误，恳请广大专家、读者批评指正，在此深表感谢。

编者

2017年5月

目 录

Unit 1 Basic Mechanical Technology ·················· 1
  Part I Technical and Practical Reading ················ 1
    Passage A Lathe Accessories (I) ·········· 1
    Passage B Lathe Accessories (II) ········ 8
  Part II Glance at Conventional Machine Tool Structures ······· 13
  Part III Simulated Writing ················ 15
    Section A Match Your Skill ············· 15
    Section B Have a Try ··················· 16
  Part IV Broaden Your Horizon—Practical Activity ·············· 18

Unit 2 Metal Materials and Metal Forming ················ 20
  Part I Technical and Practical Reading ················ 20
    Passage A Metal Materials ············· 20
    Passage B Sheet Metal Forming ······· 27
  Part II Glance at Conventional Machine Tool Structures ······· 33
  Part III Simulated Writing ················ 35
    Section A Match Your Skill ············· 35
    Section B Have a Try ··················· 37
  Part IV Broaden Your Horizon—Practical Activity ·············· 38

Unit 3 Machining Operations and Turning Machines ················ 41
  Part I Technical and Practical Reading ················ 41
    Passage A Machining Operations ······ 41
    Passage B Turning Machines ············· 46
  Part II Glance at Conventional Machine Tool Structures ········ 51
  Part III Simulated Writing ················ 53
    Section A Match Your Skill ············· 53
    Section B Have a Try ··················· 54
  Part IV Broaden Your Horizon—Practical Activity ·············· 55

Unit 4 Hydraulic Machinery and Forging Equipment ·············· 60
  Part I Technical and Practical Reading ················ 60
    Passage A Hydraulic Machinery ·········· 60
    Passage B Forging Equipment ············· 66
  Part II Glance at Conventional Machine Tool Structures ········ 72
  Part III Simulated Writing ················ 74
    Section A Match Your Skill ············· 74
    Section B Have a Try ··················· 76
  Part IV Broaden Your Horizon—Practical —Activity ················ 77

Unit 5 Introduction to CNC Machine and CAM Design ···················· 79
  Part I Technical and Practical Reading ················ 79
    Passage A Basics of Computer Numerical Control ············· 79
    Passage B Designing Parts with CAM Alibre ····················· 85

Part II Glance at Conventional Machine Tool Structures ……… 90
Part III Simulated Writing …………… 92
 Section A Match Your Skill ……… 92
 Section B Have a Try ……………… 93
Part IV Broaden Your Horizon—Practical Activity …………… 95

Unit 6 Engineering Drawings ……… 98

Part I Technical and Practical Reading ……………………… 98
 Passage A Technical Drawing (I) ……… 98
 Passage B Technical Drawing (II) …… 103
Part II Glance at Technical Drawing Instruments ………………… 108
Part III Simulated Writing …………… 109
 Section A Match Your Skill ……… 109
 Section B Have a Try ……………… 110
Part IV Broaden Your Horizon—Practical Activity …………… 111

Unit 7 Electronic Components and Circuit ……………… 114

Part I Technical and Practical Reading ……………………… 114
 Passage A Electronic Components and Symbols ………………… 114
 Passage B Basic Circuit Concepts …… 119
Part II Glance at Electronic Component Structures ………… 123
Part III Simulated Writing …………… 124
 Section A Match Your Skill ……… 124
 Section B Have a Try ……………… 125
Part IV Broaden Your Horizon—Practical Activity …………… 127

Unit 8 Single Chip Microprocessor ……………… 129

Part I Technical and Practical Reading ……………………… 129
 Passage A Introduction to Single Chip Microprocessor and Its Circuit ……………………… 129
 Passage B Introduction to MCS51 series Single Chip Microprocessor ………… 134
Part II Glance at Single Chip Microprocessor Structures …… 138
Part III Simulated Writing …………… 140
 Section A Match Your Skill ……… 140
 Section B Have a Try ……………… 141
Part IV Broaden Your Horizon—Practical Activity …………… 143

Unit 9 Introduction to Motors ……… 146

Part I Technical and Practical Reading ……………………… 146
 Passage A The Motor Basics (I) ……… 146
 Passage B The Motor Basics (II) …… 151
Part II Glance at Automatic Control Structures ………………… 156
Part III Simulated Writing …………… 157
 Section A Match Your Skill ……… 157
 Section B Have a Try ……………… 159
Part IV Broaden Your Horizon—Practical Activity …………… 160

Unit 10 Introduction to Programmable Logic Controller …………… 164

Part I Technical and Practical Reading ……………………… 164
 Passage A Programmable Logic Controller (PLC) ………… 164
 Passage B Connection of Programmable Logic Controllers ………… 169
Part II Glance at Programmable Logic Control ………………… 175
Part III Simulated Writing …………… 177

Passage A	Match Your Skill	177
Section B	Have a Try	178
Part IV	Broaden Your Horizon—Practical Activity	179

Glossary ·············· 182

Phrases ·············· 201

参考译文 ·············· 210
 第1单元 机械技术基础 ·············· 210
  课文A 车床附件（Ⅰ）·············· 210
  课文B 车床附件（Ⅱ）·············· 213
 第2单元 金属材料和金属成形 ·············· 215
  课文A 金属材料 ·············· 215
  课文B 金属板成形 ·············· 219
 第3单元 加工操作与车削机床 ·············· 222
  课文A 加工操作 ·············· 222
  课文B 车削机床 ·············· 224
 第4单元 液压机械和锻压设备 ·············· 226
  课文A 液压机械 ·············· 226
  课文B 锻压设备 ·············· 228
 第5单元 计算机数控机床和计算机辅助制造设计简介 ·············· 230
  课文A 计算机数控的基础 ·············· 230
  课文B 用Alibre计算机辅助制造软件设计零件 ·············· 233
 第6单元 机械制图 ·············· 236
  课文A 工程图（Ⅰ）·············· 236
  课文B 工程图（Ⅱ）·············· 238
 第7单元 电器元件与电路 ·············· 241
  课文A 电器元件及符号 ·············· 241
  课文B 电路的基本概念 ·············· 242
 第8单元 单片机 ·············· 244
  课文A 单片机及其电路简介 ·············· 244
  课文B MCS51系列单片机简介 ·············· 245
 第9单元 电机介绍 ·············· 247
  课文A 电机基础（Ⅰ）·············· 247
  课文B 电机基础（Ⅱ）·············· 250
 第10单元 可编程逻辑控制器简介 ·············· 252
  课文A 可编程逻辑控制器 ·············· 252
  课文B 可编程逻辑控制器的连接 ·············· 254

参考文献 ·············· 259

Unit 1
Basic Mechanical Technology

| Part I   Technical and Practical Reading |

Passage A   Lathe Accessories (I)

Brass Round Stock

Brass is a nice material to work with, though somewhat expensive compared with aluminum or steel. It can add a nice touch of contrasting color to a project that will be displayed. The alloy most often used for home shop work is 360 (Figure 1-1).

Center Drills

Center drills are stiff, stubby little drills used to start holes in the end of work-piece. If you try to drill a hole in a work-piece without using a center drill you will find that the drill will most likely wobble off center and not drill straight into the work-piece.

Standard drilling practice is to first make a facing cut on the end of the work-piece, then drill a starting hole using a center drill and then drill the hole to the required depth with a standard drill (Figure 1-2).

Chip Brushes

Chip brushes are inexpensive paint brushes that are handy for all kinds of uses around the shop. We're not sure where the name originated, but we don't think it had anything to do with removing chips; nevertheless this is one of the uses these brushes excel at. They are also ideal to clean the packing grease off a new lathe or mill. Incidentally, experienced machinists will tell you always to use a brush, rather than compressed air, to clean the chips from machine tools as compressed air will drive the chips deep into the recesses of the machine. Chip brushes and a shop vac are the preferred way to clean up chips (Figure 1-3).

Figure 1-1　Brass Round Stock

Figure 1-2　Center Drills

Digital Caliper

Digital calipers are used just like dial calipers for making inside and outside measurements accurate to one thousandth of an inch but have a direct LCD digital readout. On a dial caliper you first read the major dimension to the nearest tenth of an inch from the slide and mentally add to that the minor dimension from the dial to the nearest thousandth. This becomes second nature after a while, but still introduces opportunities to make a mistake. A digital caliper reads out the full dimension on the display, so is pretty foolproof, as long as it is properly zeroed. You can also switch between metric and inch modes as needed (Figure 1-4).

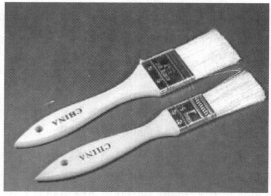

Figure 1-3　Chip Brushes

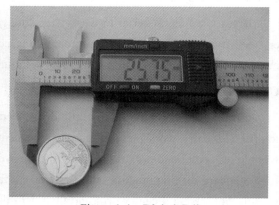

Figure 1-4　Digital Caliper

Drill Rod

Drill rod is a steel alloy with a shiny silvery color and good machining properties. Unlike other raw materials, which may vary from the nominal diameter by +0.010 or more, drill rod is surface ground to within about 0.001 of the nominal diameter. steel, drill rod is moderately resistant to rust — more, at least, that great for applications such as shafts and axles.

A useful property of drill rod is that it can easily be hardened by heating to a red then quenching in oil or water. Thus treated, the metal is hard enough to use for tools such punches (Figure 1-5).

Drill Sets

Drilling is one of the most commonly performed operations on the lathe, so you will need a good collection of decent-quality drills. When you buy your lathe, don't forget to order a tailstock chuck and arbor to hold the drills.

Poor quality drills are easy to find, but they are truly a waste of money. That's not to say that you need to buy top quality industrial drills. It's not always easy to tell a good drill bit from a poor one just by looking and, of course, its even harder if all you have is a picture on a web site or in a catalog. Generally, though, the lowest priced drill sets are the ones to stay away from (Figure 1-6).

Figure 1-5   Drill Rod

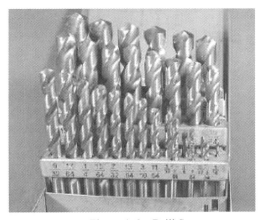

Figure 1-6   Drill Sets

Faceplate

A faceplate is a handy accessory for turning odd-shaped work that cannot easily be held in a chuck. While it's not too difficult to make your own faceplate from steel or aluminum, at this price it's hardly worth the effort.

After mounting the faceplate on the spindle, it is standard practice for faceplates to take a light one-time facing cut to ensure that the face of the plate is square with the lathe. Cast iron dust is very harmful to breathe, so I strongly recommend wearing a dust mask during this operation and until you have vacuumed up the resulting dust.

When using a faceplate, always ensure that the work is securely clamped down and balanced by some offsetting piece of metal, if necessary (Figure 1-7).

...ate thicknesses, typically from 0.001 up to ... about 20 to 40 separate strips in a set, joined ... the end of each gage. Each gage has the thickness ... example below. All of the leaves fold up into the handle ... they're not in use.

Figure 1-7　Faceplate

Figure 1-8　Feeler Gages

While their intended purpose is to measure the gap between two surfaces, such as the electrodes of spark plug.

Follower Rest

A follower rest is similar to a steady rest, but is attached to and travels with the carriage to provide a moving support for the work behind the cutting tool. This is very handy when trying to turn limber work which would otherwise bow out away from the tool. If you have ever wondered about the two screw holes on the left edge of the carriage, now you know what they are for — they are the mounting holes for the follower rest (Figure 1-9).

Figure 1-9　Follower Rest

Notes

1. Center drills are stiff, stubby little drills used to start holes in the end of work-piece. 本句是个简单句。句中 used to start…是后置定语，修饰 stubby little drills。全句可译为：中心钻是坚硬的、粗而短的小型钻头，用于在工件端面开孔。

2. Digital calipers are used just like dial calipers for making inside and outside measurements accurate to one thousandth of an inch but have a direct LCD digital readout. 本句稍长，但它是一个并列句。句中"but"为连词，译为"而且"。全句可译为：数显卡尺就像带表卡尺一样，

被用来实现千分之一英寸的内外精确测量，而且带有一个液晶显示器。

3. After mounting the faceplate on the spindle, it is standard practice for faceplates to take a light one-time facing cut to ensure that the face of the plate is square with the lathe. 本句是一个复合句，After…引导出介词短语，主语为 to take…，而 to ensure that…又引导出一个状语从句。全句可译为：在把花盘安装在主轴上后，常规作法是对断面进行一次性的轻切削以确保花盘表面与车床主轴垂直。

New Words

lathe [leɪð] n. 车床
accessory [ækˈsesəri] n. 附件
brass [brɑːs] n. 黄铜
stock [stɔk] n. 棒料
aluminum [əˈljuːminəm] n. 铝
alloy [ˈælɔi] n. 合金
drill [dril] n. 钻头
             v. 钻
stubby [ˈstʌbi] a. 粗而短的
wobble [ˈwɔbl] v. 摆动
chip [tʃip] n. 铁屑
grease [griːs] n. 润滑脂
mill [mil] n. 铣床
incidentally [insiˈdentəli] ad. 顺便说一句
machinist [məˈʃiːnist] n. 机械师
compress [kəmˈpres] v. 压缩
recess [riˈses] n. 凹窝处，沟槽
preferred [priˈfəːd] a. 优先的，首选的
digital [ˈdidʒitl] a. 数字的
caliper [ˈkæli pə] n. 卡尺
dial [ˈdaiəl] n. 刻度表
measurement [ˈmeʒəmənt] n. 测量
accurate [ˈækjurit] a. 精确的
readout [ˈriːdaut] n. 读数，读数器
major [ˈmeidʒə] a. 大的
dimension [diˈmenʃən] n. 尺寸
slide [slaid] n. 滑尺
minor [ˈmainə] a. 小的
foolproof [ˈfuːlpruːf] a. 不会出错的
metric [ˈmetrik] n. 米制

rod [rɔd]　n. 杆
silvery ['silvəri]　a. 银色的，似银的
nominal ['nɔminl]　a. 公称的，基本的
diameter [dai'æmitə]　n. 直径
grind [graind]　v. 磨削
stainless ['steinlis]　a. 不生锈的
resistant [ri'zistənt]　a. 抵抗的，阻止的
rust [rʌst]　n. 锈
carbon ['kɑ:bən]　n. 碳
shaft [ʃɑ:ft]　n. 传动轴
axle ['æksl]　n. 车轴，轮轴
quench [kwentʃ]　v. （淬火）冷却
punch [pʌntʃ]　n. 冲头，冲床
decent-quality ['di:snt'kwɔliti]　a. 优质的
tailstock ['teilstɔk]　n. 尾座
chuck [tʃʌk]　n. 夹具
arbor ['ɑ:bə]　n. 刀杆
faceplate ['feispleit]　n. （车床）花盘
odd-shaped [ɔdʃeipt]　a. 奇特形状的，不规则的
mount [maunt]　v. 安装
spindle ['spindl]　n. 主轴
cast [kɑ:st]　n. 铸造
mask [mɑ:sk]　n. 口罩
vacuum ['vækjuəm]　v. 用吸尘器吸
　　　　　　　　　n. 真空
securely [si'kjuəli]　ad. 安全地
clamp [klæmp]　v. 夹紧
offset ['ɔ:fset]　v. 偏置
gage [geidʒ]　n. 量规
strip [strip]　n. 金属片
bolt [bəult]　n. 螺栓
leave [li:v]　n. 金属片
fold [fəuld]　v. 折
electrode [i'lektrəud]　n. 电极板，电极棒
spark [spɑ:k]　n. 火花
plug [plʌg]　n. 塞子
attach [ə'tætʃ]　v. 固定
carriage ['kæridʒ]　n. 滑板
limber ['limbə]　a. 柔性的，易弯曲的

screw [skru:] n. 螺钉

Phrases and Expressions

contrasting color 对比色
center drill 中心钻
chip brush 清屑刷
be handy for 便于
excel at 擅长
packing grease 密封润滑脂
digital caliper 数显卡尺
dial caliper 刻度卡尺
LCD (Liquid Crystal Display) 液晶显示器
drill rod 钻杆
machining property 加工性能
raw material 原材料
carbon steel 碳素钢
a good collection of 大量的
standard practice 常规作法
be square with 与……垂直
cast iron 铸铁
feeler gage 测隙规，塞尺，厚薄规
intended purpose 预期作用
steady rest 固定支架
spark plug 火花塞
follower rest 跟刀架
bow out away from the tool 让刀（工件在切削力作用下产生的躲刀现象）

EXERCISE 1

The following is set of symbols denoting Working Safety. Choose the best symbol according to the information given below.

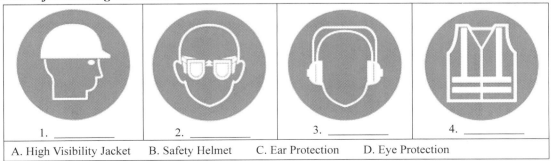

1. _____   2. _____   3. _____   4. _____
A. High Visibility Jacket   B. Safety Helmet   C. Ear Protection   D. Eye Protection

EXERCISE 2

Translate the following phrases into Chinese or English.

1. center drill   _____
2. _____   密封润滑脂
3. machining property   _____
4. _____   碳素钢
5. cast iron   _____
6. _____   测隙规
7. steady rest   _____
8. _____   跟刀架

Passage B   Lathe Accessories (II)

Live Center

Centers are often used in the tailstock to support the end of a relatively long and limber work-piece. The drawback of a "dead" center is that the center does not rotate, while the work-piece that it supports does, leading to friction and possibly overheating. By contrast, the tip of a "live" center rotates freely in bearings, and rotates with the work-piece so that the friction is greatly reduced (Figure 1-10).

Medium Density Fiberboard (MDF)

MDF is not really an accessory, but it is a handy, inexpensive, lightweight and easily workable material to have around the shop. It is a synthetic material, but with different properties. It is available in thicknesses up to at least 3/4" and is smooth on both faces. It cuts easily using a saber saw, circular saw, band saw, table saw or radial arm saw. One disadvantage is that it tends to absorb moisture, but this propensity can be greatly reduced by spray-painting the surfaces (Figure 1-11).

Figure 1-10   Live Centers   Figure 1-11   Medium Density Fiberboard

Unit 1 Basic Mechanical Technology

Milling Attachment

A milling attachment became available as a standard accessory for the mini "–" lathe. Before this accessory became available, many lathe owners made their own versions based on the one made by Varmint Al. Varmint Al's uses a standard milling vise while the one shown below uses socket head screws to hold the work-piece between the jaws. While by no means a substitute for mini-mill, this accessory is an inexpensive way to add limited milling capability to your mini-lathe while you save up for a mill (Figure 1-12).

Figure 1-12   Milling Attachment

T-handle Metric Hex Wrench Set

These aren't really essential, but they are so handy and so cheap that you will kick yourself if you live without them and then later try them. I use these for nearly all the hex head screws on the lathe (Figure 1-13).

Tailstock Chuck and Arbor

Drilling is a fundamental lathe operation and you will need a chuck and a #2 Morse Taper arbor to do it. The arbor has a thread or Jacobs taper on one end (to mate with the chuck) and a #2MT on the other end to mate with the tailstock cylinder.

To remove the arbor, place the chuck in your bench vise with the jaws open just a little wider than the arbor diameter (not clamping the arbor). Open the jaws up and use a short piece of round stock or a drift pin to drive the arbor out of the back of the chuck. This should only require a fairly light tap of the hammer. It's a good idea to position a rag underneath the arbor to catch it so that it does not get dinged up by falling to the floor (Figure 1-14).

Figure 1-13   T-handle Metric Hex Wrench Set

Figure 1-14   Tailstock Chuck and Arbor

Tool Bits

While I recommend that you learn to grind your own tool bits from Tool Blanks, you may want to have a set of pre-ground High Speed Steel (HSS) tool bits. Not only will they get you off to an immediate start, they'll serve as good examples of the shapes you want to attain when you grind your own. Here's a very nice set from Walden Specialties. The set includes a 60 threading tool that can be difficult to grind by hand (Figure 1-15).

Quick Change Tool-post

A quick change tool post (QCTP) is an accessory that we strongly recommend: get one as soon as your budget allows. Why? It will save you hours of time and lower your blood pressure by several points since you'll no longer have to stack shims under the tool bit to get it to match the height of the lathe centerline. With a QCTP, each cutting tool has its own dedicated holder. Each holder has a locking height adjustment. A few quick trials and adjustments and you lock the tool height to the perfect setting. You won't need to adjust it until you sharpen the tool, which may lower the tip by a few thousandths (Figure 1-16).

Figure l-15　Toollls Bits

Figure l-16　Quick Change Tool Post

Wet/Dry Sandpaper

Wet/Dry sandpaper is very handy for putting a very fine shiny finish on metal work-pieces in the lathe. In that application it is generally used dry, but when used wet on a flat backing surface it does a great job of smoothing flat metal workpieces from milling operations (Figure 1-17).

White Lithium Grease

White lithium grease is useful for lubricating just about any of the moving parts of the lathe (Use light oil on the ways, though, not greases) (Figure 1-18).

Figure 1-17　Wet/Dry Sandpaper

Figure 1-18　White Lithium Grease

New Words

center ['sentə]　n. 顶尖

Unit 1 Basic Mechanical Technology

drawback ['drɔːˌbæk]　n. 弊端
rotate [rəu'teit]　v. 旋转
friction ['frikʃən]　n. 摩擦
bearing ['bɛəriŋ]　n. 轴承
density ['densiti]　n. 密度
synthetic [sin'θetic]　a. 合成的
saw [sɔː]　n. 锯
circular ['səːkjulə]　a. 圆形的
band [bænd]　n. 带状
radial ['reidjəl]　a. 径向的
moisture ['mɔistʃə]　n. 水分
propensity [prə'pensiti]　n. 倾向性
attachment [ə'tætʃmənt]　n. 夹具
version ['vəːʃən]　n. 版本，型号
vise [vais]　n. 虎钳
jaw [dʒɔː]　n. 夹具
substitute ['sʌbstitjuːt]　n. 替代品
hexagonal [hek'sægənəl]　a. 六角形的
wrench [rentʃ]　n. 扳手
thread [θred]　n. 螺纹
taper ['teipə]　n. 锥度
mate [meit]　v. 匹配
cylinder ['silində]　n. 汽缸
bench [bentʃ]　n. 工作台
hammer ['hæmə]　n. 锤子
rag [ræg]　n. 地脚螺栓
ding [diŋ]　v. 发出声响
attain [ə'tein]　v. 获得，达到
stack [stæk]　v. 堆
shim [ʃim]　n. 垫片
centerline ['sentəlain]　n. 中线
dedicated ['dedɪkeɪtɪd]　a. 专用的
holder ['həuldə]　n. 刀架
adjustment [ə'dʒʌstmənt]　n. 调整
trial ['traiəl]　n. 试刀
sandpaper ['sændpeɪpə(r)]　n. 砂纸
finish ['finiʃ]　n. 抛光度
lithium ['liθiəm]　n. 锂
lubricate ['luːbrikeit]　v. 润滑

Phrases and Expressions

live center　活动顶尖
by contrast　相反
Medium Density Fiberboard (MDF)　中密度纤维板
saber saw　军刀形电动手锯
circular saw　圆盘锯
bandsaw　带锯
table saw　台锯
radial arm saw　手拉锯
tend to　易于
be based on　基于
socket head screw　沉头螺钉
by no means　绝不
T-handle Metric Hex Wrench Set　T型手柄公制成套六角扳手
Morse Taper　莫氏锥度
Jacobs Taper　贾克布锥度
drift pin　冲头
tool bit　刀头
Tool Blank　刀头
High Speed Steel (HSS)　高速钢
Walden Specialties　瓦尔登专卖店
threading tool　螺纹车刀
by hand　用手工
quick change tool post (QCTP)　快换刀架
white lithium grease　白色锂基润滑脂
light oil　轻质油

EXERCISE 3

Choose the best lathe accessory according to the information given.

1. 	2.	3.	4. 	5.	6.

A. Metric Hex Key Set　　　　　　B. Live Center　　　　　　C. Tool Bits
D. Feeler Gages　　　　　　　　　E. Machine Vice　　　　　　F. Digital Caliper

Unit 1 Basic Mechanical Technology

EXERCISE 4

Abbreviations are very useful in practical work. Read them and then translate them into corresponding Chinese terms.

1. A.C.	Alternating Current	_____
2. ACRE	Automatic Checkout and Readiness Equipment	_____
3. AE	Absolute Error	_____
4. AED	Automated Engineering Design	_____
5. A/M	Auto and Manual	_____
6. ARC	Automatic Ratio Control	_____
7. BAP	Basic Assembler Program	_____
8. BG	Bevel Gear	_____

| Part II   Glance at Conventional Machine Tool Structures |

The following is the structure of conventional machine tool .

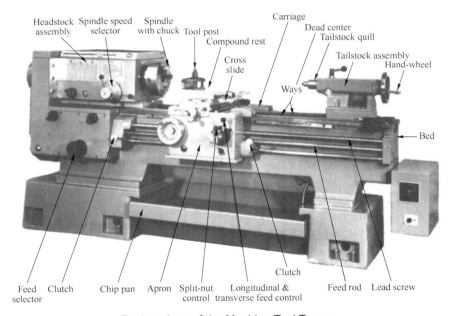

Explanations of the Machine Tool Terms

No.	Name	Explanation
1	Spindle speed selector	主轴速度转换开关
2	Headstock assembly	主轴箱
3	Spindle with chuck	附有夹具主轴
4	Tool post	刀架
5	Compound rest	复式刀架

续表

No.	Name	Explanation
6	Cross slide	横向拖板
7	Carriage	滑鞍，滑座
8	Ways	导轨
9	Dead center	死顶尖
10	Tailstock quill	尾架顶尖套筒
11	Tailstock assembly	尾架
12	Hand-wheel	手轮
13	Bed	底座，床身
14	Lead screw	丝杠
15	Feed rod	进刀杠，光杠
16	Clutch	离合器
17	Longitudinal & transverse feed control	纵向和横向进给控制
18	Split-nut control	开合螺母控制
19	Apron	溜板箱，进给箱
20	Chip pan	承屑盘
21	Feed selector	进给选择开关

EXERCISE 5

The following is the conventional machine tool. You are required to choose the suitable words or phrases given below.

| Special Purpose Lathe | Pipe Screw-cutting Lathe | Vertical Lathe |
| Double Turret Lathe | Horizontal Lathe | Automatic Lathe |

1. _____ （双刀塔车床）

Common Lathe（普通车床）

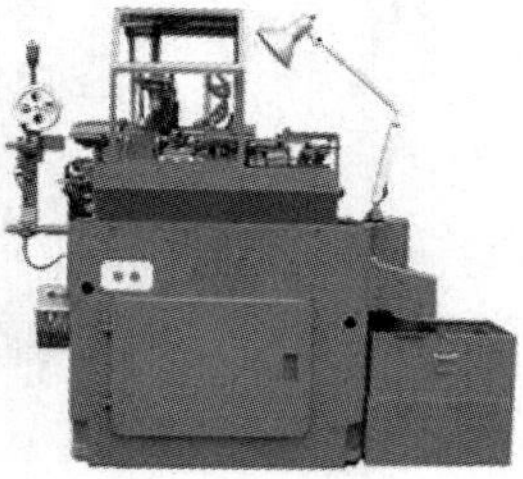

2. _____ （自动车床）

3. _____ （立式车床）

续表

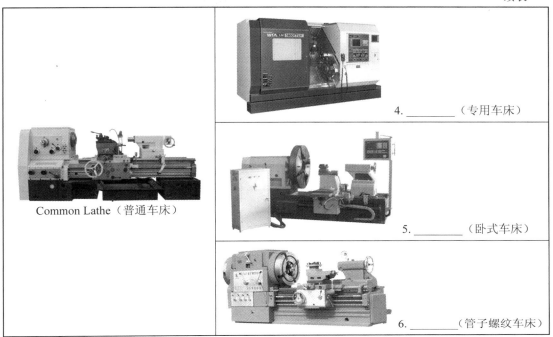

Common Lathe（普通车床）

4. _____（专用车床）

5. _____（卧式车床）

6. _____（管子螺纹车床）

| Part III Simulated Writing |

Section A Match Your Skill

The following is a lathe accessory, and you can understand the name of Tool Bits.

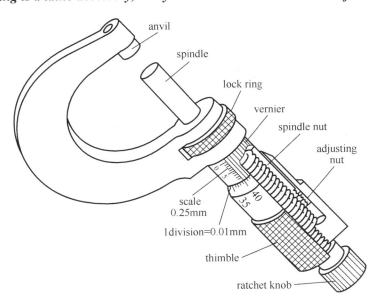

Explanations of Terms

No.	Name	Explanation
1	Anvil	固定爪
2	Spindle	伸缩爪（轴）
3	Lock ring	防松环
4	Vernier	游标尺
5	Spindle nut	轴螺母
6	Adjusting nut	调节螺母
7	Scale	刻度
8	Thimble	外套管
9	Ratchet knob	棘轮旋钮

EXERCISE 6

Match the words or phrases on the left with their meanings on the right.

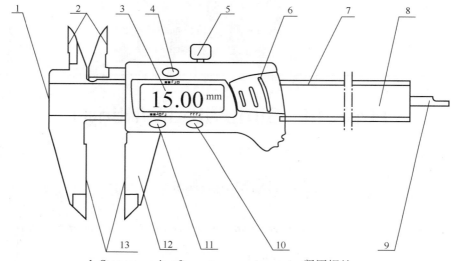

1. Step measuring face   A. 紧固螺丝
2. Inside measuring face   B. 校准贴纸
3. LCD Display Screen   C. 内径测量面
4. inch/mm key   D. 尺身
5. Locking screw   E. 阶梯测量面
6. Battery cover   F. 开/关键
7. Body   G. 公英制转换键
8. Calibration sticker   H. 液晶显示屏
9. Depth bar   I. 外径测量面
10. Zero-setting key   J. 滑尺
11. On/Off key   K. 电池盖
12. Slider   L. 清零键
13. Outside measuring face   M. 测深杆

Section B Have a Try

This section will help you to understand several forms of machining.

External Operations

Turning — A turning operation in which a single-point tool moves axially, along the side of the work "-" piece, removing material to form different features, including steps, tapers, chamfers, and contours. These features are typically machined at a small radial depth of cut and multiple passes are made until the end diameter is reached. For a finish operation, the cutting feed is calculated based on the desired surface roughness and the tool nose radius.

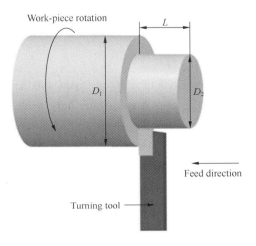

EXERCISE 7

This section is to test your ability to identify different operations.

Contours	Chamfers	Tapers

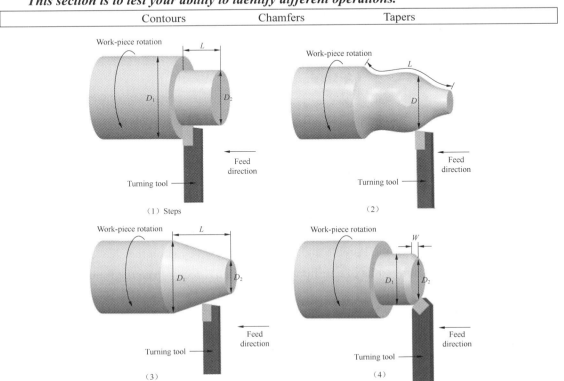

A single-point turning tool moves axially, along the side of the work-piece, removing material to form different features, including (1) steps,…

(1) Steps   (2) _____   (3) _____   (4) _____

Part IV   Broaden Your Horizon— Practical Activity

The micrometer screw gauge

The micrometer screw gauge is used to measure even smaller dimensions than the vernier caliper. The micrometer screw gauge also uses an auxiliary scale (measuring hundredths of a millimeter) which is marked on a rotary thimble. Basically it is a screw with an accurately constant pitch. The micrometers in our laboratory have a pitch of 0.50 mm. The rotating thimble is subdivided into 50 equal divisions. The thimble passes through a frame that carries a millimetre scale graduated to 0.5 mm. The jaws can be adjusted by rotating the thimble using the small ratchet knob. The thimble must be rotated through two revolutions to open the jaws by 1 mm. Here is a useful applet to learn how to use the micrometer screw gauge (Figure 1-19).

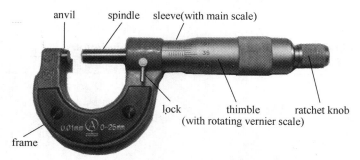

Figure 1-19   The micrometer screw gauge

In order to measure an object, the object is placed between the jaws and the thimble is rotated using the ratchet until the object is secured. Note that the ratchet knob must be used to secure the object firmly between the jaws, otherwise the instrument could be damaged or give an inconsistent reading.

Note that an additional half scale division (0.5 mm) must be included if the mark below the main scale is visible between the thimble and the main scale division on the sleeve. The remaining two significant figures (hundredths of a millimeter) are taken directly from the thimble opposite the main scale.

In Figure 1-20 the last graduation visible to the left of the thimble is 7 mm and the thimble lines up with the main scale at 38 hundredths of a millimeter (0.38 mm); therefore the reading is 7.38 mm.

In Figure 1-21 the last graduation visible to the left of the thimble is 7.5 mm; therefore the reading is 7.5 mm plus the thimble reading of 0.22 mm, giving 7.72 mm.

Figure 1-20   The reading is 7.38mm

Figure 1-21   The reading is 7.27mm

In Figure 1-22 the main scale reading is 3 mm while the reading on the drum is 0.46 mm; therefore, the reading is 3.46 mm.

In Figure 1-23 the 0.5 mm division is visible below the main scale; therefore the reading is 3.5 mm + 0.06 mm = 3.56 mm.

Figure 1-22   The reading is 3.46mm

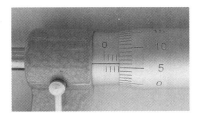

Figure 1-23   The reading is 3.56mm

Try the following bg yourself

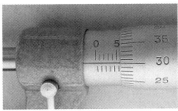

Figure 1-24

Figure 1-25

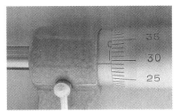

Figure 1-26

Answer: Figure 1-24   5.80mm; Figure 1-25   3.09mm; Figure 1-26   0.29mm。

Unit 2
Metal Materials and Metal Forming

| Part I   Technical and Practical Reading |

Passage A   Metal Materials

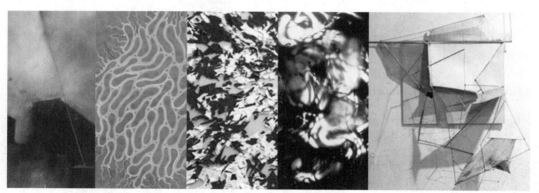

Almost 75% of all elements are metals. Metals are used in electronics for wires and in cookware for pots and pans because they conduct electricity and heat well. Most metals are malleable and ductile and are, in general, heavier than the other elemental substances. Two or more metals can be alloyed to create materials with properties that do not exist in a pure metal.

All metals can be classified as either ferrous or non-ferrous. Ferrous metals contain iron and non-ferrous metals do not. All ferrous metals are magnetic and have poor corrosion resistance while non-ferrous metals are typically non-magnetic and have more corrosion resistance. An overview of the most common ferrous and non-ferrous metals is shown below.

Ferrous Metals

(1)

Material name: Low Carbon Steels (Figure 2-1)
Composition: Up to 0.30% carbon
Properties: Good formability, good weldability, low cost
Applications: 0.1%～0.2% carbon: Chains, stampings, rivets, nails, wire, pipe, and so on

0.2%～0.3% carbon: Machine and structural parts

(2)

Material name: Medium Carbon Steels (Figure 2-2)

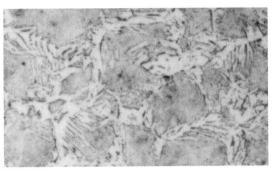

Figure2-1　Low Carbon Steels　　　　Figure2-2　Medium Carbon Steels

Composition: 0.30% to 0.80% carbon

Properties: A good balance of properties, fair formability

Applications: 0.3%～0.4% carbon: Lead screws, gears, worms, spindles, shafts, and machine parts

　　　　　　0.4%～0.5% carbon: Crankshafts, gears, axles, mandrels, tool shanks, and heat-treated machine parts

　　　　　　0.6%～0.8% carbon: Drop hammer dies, set screws, screwdrivers, and arbors

　　　　　　0.7%～0.8% carbon: Tough and hard steel. Anvil faces, band saws, hammers, wrenches, and cable wire

(3)

Material name: High Carbon Steels (Figure 2-3)

Composition: 0.80%～2.0% Carbon

Properties: Low toughness, formability, and weld-ability, high hardness and wear resistance, fair formability

Applications: 0.8%～0.9% carbon: Punches for metal, rock drills, shear blades, cold chisels, rivet sets, and many hand tools

　　　　　　0.9%～1.0% carbon: Used for hardness and high tensile strength, like springs and cutting tools

　　　　　　1.0%～1.2% carbon: Drills, taps, milling cutters, knives, cold cutting dies, wood working tools

　　　　　　1.2%～1.3% carbon: Files, reamers, knives, tools for cutting wood and brass

　　　　　　1.3%～1.4% carbon: Used where a keen cutting edge is necessary (razors, saws, etc.) and where wear resistance is important

(4)

Material name: Stainless Steel (Figure 2-4)

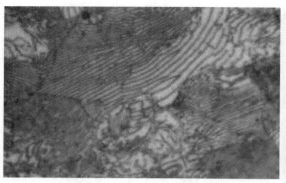

Figure 2-3　High Carbon Steels

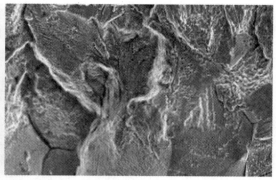

Figure 2-4　Stainless Steels

Composition: Stainless steel is a family of corrosion resistant steels. They contain at least 10.5% chromium. The Chromium in the alloy forms a self-healing protective clear oxide layer. This oxide layer gives stainless steels their corrosion resistance

Properties: Good corrosion resistance, appearance, and mechanical properties

Applications: 11.5% of chromium: Used in cookware, cutlery, and kitchen utensils widely. These are along with hardware supplies, industrial equipments, structural buildings, automotive, and aerospace industries

Non-Ferrous Metals

(5)

Material name: Aluminum/Aluminum Alloys (Figure 2-5)

Composition: Pure metal; easily alloyed with small amounts of copper, manganese, silicon, magnesium, and other elements

Properties: Low density, good electrical conductivity (approx. 60% of copper), nonmagnetic, noncombustible, ductile, malleable, corrosion resistance; easily formed, machined, or cast

Applications: Window frames, aircraft parts, automotive parts, kitchenware

(6)

Material name: Brass (Figure 2-6)

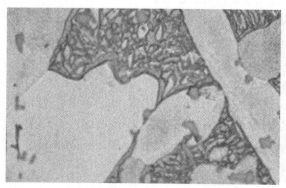

Figure 2-5　Aluminum alloys

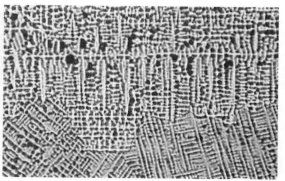

Figure 2-6　Brass

Composition: Alloy of copper and zinc; 65% to 35% is the common ratio
Properties: Reasonable hardness; casts, forms, and machines well; good electrical conductivity and acoustic properties
Applications: Parts for electrical fittings, valves, forgings, ornaments, musical instruments

(7)

Material name: Copper (Figure 2-7)
Composition: Pure metal
Properties: Excellent ductility, thermal and electrical conductivity
Applications: Electrical wiring, tubing, kettles, bowls, pipes, printed circuit boards

(8)

Material name: Magnesium/Magnesium Alloys (Figure 2-8)

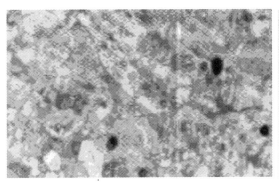

Figure 2-7   Copper   Figure 2-8   Magnesium Alloys

Composition: Pure metal; used as an alloy element for aluminum, lead, zinc, and other nonferrous alloys; alloyed with aluminum to improve the mechanical, fabrication, and welding characteristics
Properties: Lightest metallic material (density of about 2/3 of that of aluminum), strong and tough, most machinable metal, good corrosion resistance, easily cast
Applications: Automobile, portable electronics, appliances, power tools, sporting goods parts, and aerospace equipments

(9)

Material name: Nickel / Nickel Alloys (Figure 2-9)
Composition: Pure metal; alloys very well with large amounts of other elements, chiefly chromium, molybdenum, and tungsten
Properties: Very good corrosion resistance (can be alloyed to extend beyond stainless steels), good high temperature and mechanical performance, fairly good conductor of heat and electricity
Applications: Used as an undercoat in decorative chromium plating and to improve corrosion resistance; applications include electronic lead wires, battery components, heat exchangers in corrosive environments

(10)

Material name: Zinc / Zinc Alloys (Figure 2-10)

Composition: Pure metal; employed to form numerous alloys with other metals. Alloys of primarily zinc with small amounts of copper, aluminum, and magnesium are useful in die-casting.

Properties: Excellent corrosion resistance, light weight, reasonable conductor of electricity

Applications: Used principally for galvanizing iron (more than 50% of metallic zinc goes into galvanizing steel), numerous automotive applications because of its light weight

Figure 2-9　Nickel Alloys

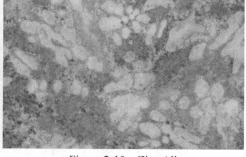

Figure 2-10　Zinc Alloys

Notes

1. Metals are used in electronics for wires and in cookware for pots and pans because they conduct electricity and heat well. 这句中，be used in "被用于…"。because 引导原因状语从句。在这个从句中，conduct 有两个宾语：electricity 和 heat。全句可译为：金属在电子行业中用来作导线，在炊具中用来作锅碗瓢盆，这是因为它们导电、导热性能都好。

2. Two or more metals can be alloyed to create materials with properties that do not exist in a pure metal. 主语为 Two or more metals。with 意为"具有"。that 引导定语从句，先行词为 properties。全句可译为：两种或更多种的金属可以形成合金材料，这些合金材料具有纯金属不存在的特性。

3. All ferrous metals are magnetic and have poor corrosion resistance while non-ferrous metals are typically non-magnetic and have more corrosion resistance. 此句中，while 用作连词，连接两个并列的句子，表示对比关系。其中，两个并列句的谓语动词又都是由 and 连接的并列谓语。该句可译为：所有的黑色金属都具有磁性和较差的耐蚀性，而有色金属通常不具有磁性，却有较强的耐蚀性。

New Words

cookware [ˈkukweə(r)]　n. 炊具

conduct [ˈkɔndʌkt]　v. 传导

malleable ['mæliəbl]   a. 可塑的
ductile ['dʌktail]   a. 易延展的，有韧性的
property ['prɔpəti]   n. 属性
ferrous ['ferəs]   a. 含铁的
magnetic [mæg'netik]   a. 有磁性的
corrosion [kə'rəuʒən]   n. 腐蚀
composition [kɔmpə'ziʃən]   n. 成分
formability [fɔ:mə'biliti]   n. 成形性
weld [weld]   n. 焊接
              v. 焊接
chain [tʃein]   n. 链条
stamping ['stæmpiŋ]   n. 冲压件
rivet ['rivit]   n. 铆钉
nail [neil]   n. 钉子
worm [wə:m]   n. 蜗杆
crankshaft ['kræŋkʃɑ:ft]   n. 曲轴
mandrel ['mændril]   n. 心轴
shank [ʃæŋk]   n. 柄
heat-treated ['hi:ttri:tid]   a. 热处理的
screwdriver ['skru:draivə]   n. 螺丝刀
anvil ['ænvil]   n. 铁砧
toughness ['tʌfnis]   n. 韧性
hardness ['hɑ:dnis]   n. 硬度
wear [weə]   n. 磨损
shear [ʃiə]   n. 剪切
blade [bleid]   n. 刀片
chisel ['tʃizl]   n. 凿子
tensile ['tensail]   a. 可拉伸的
tap [tæp]   n. 螺丝攻
file [fail]   n. 锉刀
reamer ['ri:mə]   n. 铰刀
keen [ki:n]   a. 锋利的
razor ['reizə]   n. 剃刀
chromium ['krəumjəm]   n. 铬
heal [hi:l]   v. 愈合
oxide ['ɔksaid]   n. 氧化物
layer ['leiə]   n. 层
cutlery ['kʌtləri]   n. 餐具
utensil [ju(:)'tensl]   n. 厨具

hardware [ˈhɑːdweə]　n. 五金
aerospace [ˈɛərəuspeis]　n. 航空航天
manganese [ˌmæŋɡəˈniːz]　n. 锰
silicon [ˈsilikən]　n. 硅
magnesium [mæɡˈniːzjəm]　n. 镁
conductivity [ˌkɔndʌkˈtiviti]　n. 传导性
noncombustible [ˈnɔnkəmˈbʌstəbl]　a. 不燃的
form [fɔːm]　v. 成形
frame [freim]　n. 框
kitchenware [ˈkitʃinweə(r)]　n. 厨具
zinc [ziŋk]　n. 锌
acoustic [əˈkuːstik]　a. 声的
fitting [ˈfitiŋ]　n. 配件
valve [vælv]　n. 阀门
forging [ˈfɔːdʒiŋ]　n. 锻件
ornament [ˈɔːnəmənt]　n. 饰品
thermal [ˈθəːməl]　a. 热的
tubing [ˈtjuːbiŋ]　n. 管材
lead [liːd]　n. 铅
fabrication [ˌfæbriˈkeiʃən]　n. 制造
characteristics [ˌkæriktəˈristik]　n. 特性
machinable [məˈʃiːnəbl]　a. 可加工的
appliances [əˈplaiəns]　n. 家用电器
nickel [ˈnikl]　n. 镍
molybdenum [məˈlibdinəm]　n. 钼
tungsten [ˈtʌŋstən]　n. 钨
undercoat [ˈʌndəkəut]　n. 底漆，底层镀
plating [ˈpleitiŋ]　n. 镀饰，电镀
exchanger [iksˈtʃeindʒə]　n. 交换器
corrosive [kəˈrəusiv]　a. 腐蚀的
die-casting [daiˈkɑːstiŋ]　n. 压铸件
galvanize [ˈɡælvənaiz]　v. 镀锌

Phrases and Expressions

be classified as　分类为
ferrous metal　黑色金属
non-ferrous metal　有色金属
up to　多达

lead screw 丝杠
drop hammer die 落锤模
set screw 紧定螺钉
rock drill 凿岩机
tensile strength 抗拉强度
stainless steel 不锈钢
lead wire 引线

EXERCISE 1

The following is set of symbols denoting Working Safety. Choose the best symbol according to the information given below.

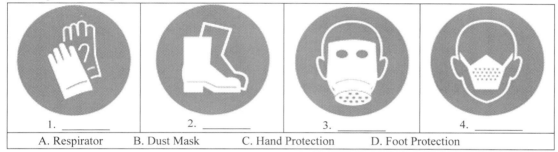

1. _____ 2. _____ 3. _____ 4. _____

A. Respirator   B. Dust Mask   C. Hand Protection   D. Foot Protection

EXERCISE 2

Translate the following phrases into Chinese or English.

1. corrosion resistance _____   2. _____ 中碳钢
3. lead screws _____   4. _____ 耐磨性
5. electrical conductivity _____ 6. _____ 铝合金
7. die-casting _____   8. _____ 印刷电路板

Passage B   Sheet Metal Forming

Bending

Bending is a metal forming process in which a force is applied to a piece of sheet metal, causing it to bend at an angle and form the desired shape. A bending operation causes deformation along one axis, but a sequence of several different operations can be performed to create a complex part. Bent parts can be quite small, such as a bracket, or up to 20 feet in length, such as a large enclosure or chassis. A bend can be characterized by several different parameters, shown in the image below (Figure 2-11).

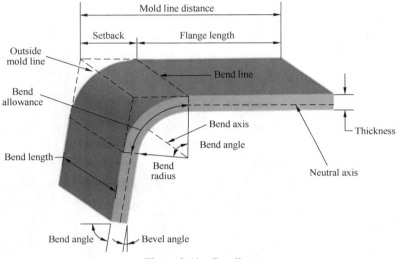

Figure 2-11   Bending

The act of bending results in both tension and compression in the sheet metal. The outside portion of the sheet will undergo tension and stretch to a greater length, while the inside portion experiences compression and shortens. The neutral axis is the boundary line inside the sheet metal, along which no tension or compression forces are present. As a result, the length of this axis remains constant. The changes in length to the outside and inside surfaces can be related to the original flat length by two parameters, the bend allowance and bend deduction, which are defined below (Figure 2-12).

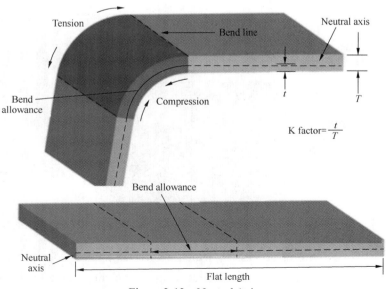

Figure 2-12   Neutral Axis

When bending a piece of sheet metal, the residual stress in the material will cause the sheet to spring-back slightly after the bending operation. Due to this elastic recovery, it is necessary to over-bend the sheet a precise amount to achieve the desired bend radius and bend angle. The final bend radius will be greater than initially formed and the final bend angle will be smaller. The ratio of the final bend angle to the initial bend angle is defined as the spring-back factor, KS. The amount of spring-back depends upon several factors, including the material, bending operation, and the initial bend angle and bend radius (Figure 2-13).

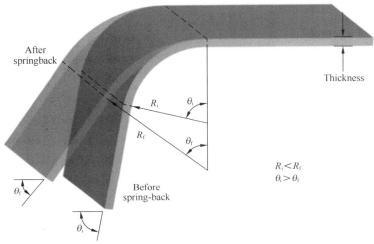

Figure 2-13 Spring-back

Deep Drawing

Deep drawing is a metal forming process in which sheet metal is stretched into the desired part shape. A tool pushes downward on the sheet metal, forcing it into a die cavity in the shape of the desired part. The tensile forces applied to the sheet cause it to plastically deform into a cup-shaped part. Deep drawn parts are characterized by a depth equal to more than half of the diameter of the part. These parts can have a variety of cross sections with straight, tapered, or even curved walls. Deep drawing is most effective with ductile metals, such as aluminum, brass, copper, and mild steel (Figure 2-14).

The deep drawing process requires a blank, a blank holder, a punch and a die. The blank is a piece of sheet metal, typically a disc or rectangle, which is pre-cut from stock material and will be

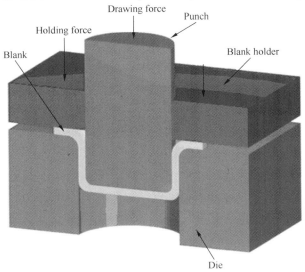

Figure 2-14 Deep Drawing

formed into the part.

The process of drawing the part sometimes occurs in a series of operations, called draw reductions. In each step, a punch forces the part into a different die, stretching the part to a greater depth each time. After a part is completely drawn, the punch and blank holder can be raised and the part removed from the die (Figure 2-15).

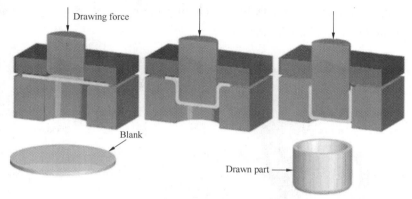

Figure 2-15   Deep Drawing Process

Stretch Forming

Stretch forming is a metal forming process in which a piece of sheet metal is stretched and bent simultaneously over a die in order to form large contoured parts. Stretch forming is performed on a stretch press, in which a piece of sheet metal is securely gripped along its edges by gripping jaws (Figure 2-16).

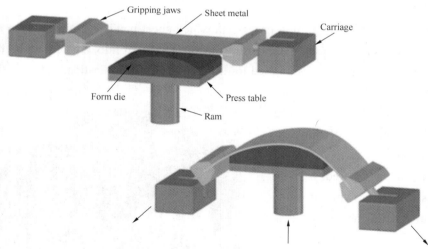

Figure 2-16   Stretch Forming

Stretch formed parts are typically large and possess large radius bends. The shapes that can be produced vary from a simple curved surface to complex non-uniform cross sections. Stretch forming is capable of shaping parts with very high accuracy and smooth surfaces. Ductile materials are preferable, the most commonly used being aluminum, steel, and titanium. Typical stretch formed parts are large curved panels such as door panels in cars or wing panels on aircraft.

The most common stretch presses are oriented vertically, in which the form die rests on a press table that can be raised into the sheet by a hydraulic ram. As the form die is driven into the sheet, which is gripped tightly at its edges, the tensile forces increase and the sheet plastically deforms into a new shape. Horizontal stretch presses mount the form die sideways on a stationary press table, while the gripping jaws pull the sheet horizontally around the form die.

New Words

sheet [ʃi:t]　n. 金属板
deformation [ˌdi:fɔ:ˈmeiʃən]　n. 变形
bracket [ˈbrækit]　n. 支架
enclosure [inˈkləuʒə]　n. 外壳
chassis [ˈʃæsi]　n. 底盘
bend [bend]　n. 弯曲成形件
characterize [ˈkæriktəraiz]　v. 以……为特性
parameter [pəˈræmitə]　n. 参数
mold [məuld]　n. 模子
allowance [əˈlauəns]　n. 余量
bevel [ˈbevəl]　n. 倾斜
neutral [ˈnju:trəl]　a. 中性的，中间的
thickness [ˈθiknis]　n. 厚度
setback [ˈsetbæk]　n. 缩入距离
flange [flændʒ]　n. 凸缘
tension [ˈtenʃən]　n. 拉张
compression [kəmˈpreʃ(ə)n]　n. 压缩
stretch [stretʃ]　v./n. 伸展
undergo [ˌʌndəˈgəu]　v. 经受
constant [ˈkɔnstənt]　a. 恒定的
factor [ˈfæktə]　n. 系数
residual [riˈzidjuəl]　a. 残余的
stress [stres]　n. 应力
spring-back [spriŋbæk]　n. 回弹
elastic [iˈlæstik]　a. 有弹性的
recovery [riˈkʌvəri]　n. 恢复
over-bend [ˈəuvəbend]　v. 过度弯曲
die [dai]　n. 模具
cavity [ˈkæviti]　n. 腔
tapered [ˈteipəd]　a. 锥形的
curved [kə:vd]　a. 弯曲的

blank [blæŋk]   n. 毛坯
disc [disk]   n. 圆盘形
rectangle ['rektæŋgl]   n. 长方形
pre-cut ['priːkʌt]   v. 预切
contoured ['kɔntuəd]   a. 波状外形的
grip [grip]   v. 夹住
non-uniform ['nɔn'juːnifɔːm]   a. 无统一形状的
titanium [tai'teinjəm]   n. 钛
panel ['pænl]   n. 板
ram [ræm]   n. 连杆
orient ['ɔːriənt]   v. 确定方向
vertically ['vəːtikəli]   ad. 沿垂直方向
stationary ['steiʃ(ə)nəri]   a. 静止的
horizontally [,hɔri'zɔntli]   ad. 沿水平方向

Phrases and Expressions

a sequence of   一系列的
outside mold line   外模线
bend allowance   弯曲余量
bend length   弯曲长度
bend angle   弯曲角
bevel angle   斜角
bend radius   弯曲半径
bend axis   弯曲轴线
bend line   弯曲线
flange length   卷边长度
boundary line   分界线
bend deduction   折弯补偿
residual stress   残余应力
die cavity   型腔
tensile force   拉力
cross section   横截面
mild steel   低碳钢
holding force   夹持力
blank holder   压料板
draw reduction   拉伸补充
stretch press   拉伸机
gripping jaw   颚形夹爪

curved surface 曲面
form die 成型模
press table 下压板

EXERCISE 3

Choose the best operation orders according to the information given.

The following is roll forming, and it is a metal forming process in which sheet metal is progressively shaped through a series of bending operations.

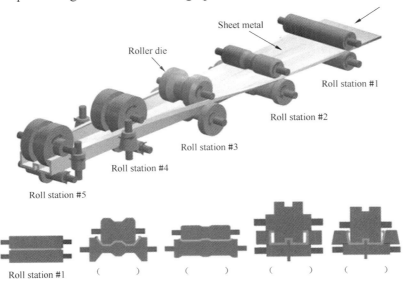

Roll station #1   (　)　　(　)　　(　)　　(　)

EXERCISE 4

Abbreviations are very useful in practical work. Read them and then translate them into corresponding Chinese terms.

1. BM	Buffer Module		_____
2. BP	Back Pressure		_____
3. CAD	Computer Aided Design		_____
4. CAM	Computer Aided Manufacturing		_____
5. CI	Circuit Interrupter		_____
6. C.I.	Cast Iron		_____
7. CIA	Alloy Cast Iron		_____
8. CL	Center Line		_____

| Part II   Glance at Conventional Machine Tool Structures |

The following is the structure of vice jaw.

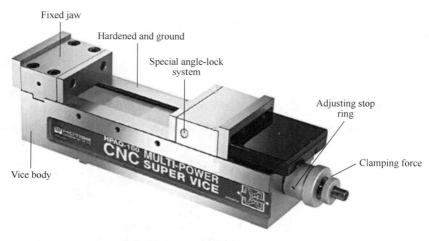

Explanations of Vice Jaw Terms

No.	Name	Explanation
1	Fixed jaw	固定钳口
2	Hardened and ground	已硬化和磨光
3	Special angle-lock system	特制的角锁紧系统
4	Vice body	虎钳体
5	Adjusting stop ring	调整止动环
6	Clamping force	夹紧力

EXERCISE 5

The following is the conventional machine tool. You are required to choose the suitable words or phrases given below.

| Single-column Milling Machine | Plano-milling Machine | Tool Milling Machine |
| Knee-and-Column Milling | Instrument milling machine | Single-arm Milling Machine |

Milling Machine（铣床）

1. _____ （升降台铣床）

2. _____ （龙门铣床）

续表

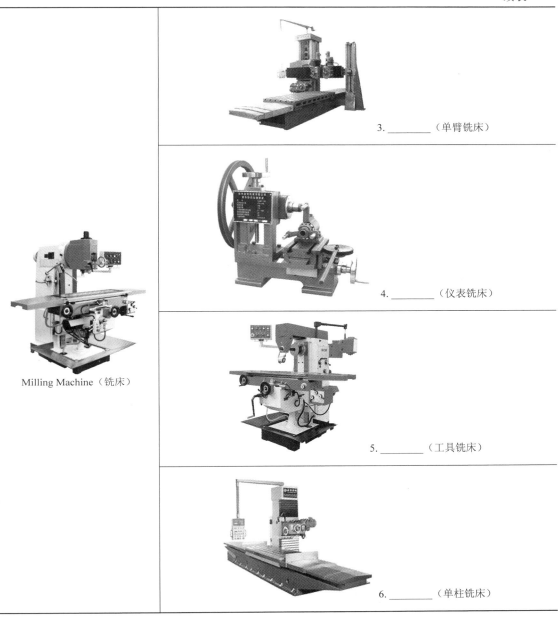

Milling Machine（铣床）

3. _____ （单臂铣床）

4. _____ （仪表铣床）

5. _____ （工具铣床）

6. _____ （单柱铣床）

Part III Simulated Writing

Section A Match Your Skill

The following is a lathe accessory, and you can understand the name of Inner Dial Caliper Gauge.

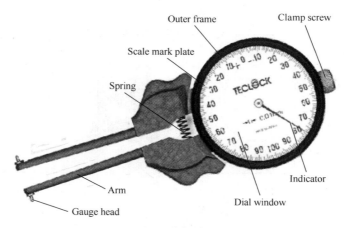

Explanations of Terms

No.	Name	Explanation
1	Outer frame	外框
2	Scale mark plate	刻度盘
3	Spring	弹簧
4	Arm	测量臂
5	Gauge head	侧头
6	Dial window	刻度盘
7	Indicator	指针
8	Clamp screw	夹紧螺丝

EXERCISE 6

Match the words or phrases on the left with their meanings on the right.

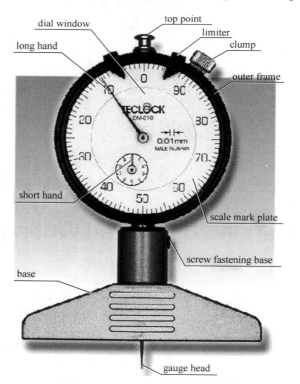

1. Dial window
2. Long hand
3. Top point
4. Limiter
5. Clump
6. Outer frame
7. Short hand
8. Base
9. Scale mark plate
10. Screw fastening base
11. Gauge head

A. 限制器
B. 短指针
C. 刻度盘
D. 刻度盘
E. 长指针
F. 底座
G. 顶针
H. 螺丝紧固座
I. 紧固件
J. 测头
K. 外框

Section B Have a Try

This section will help you to understand several forms of machining.

External Operations

Facing — A turning operation in which a single-point tool moves radially, along the end of the workpiece, removing a thin layer of material to provide a smooth flat surface. The cutting tool moves from the outer diameter to the center or inner diameter of the work-piece or can move in the opposite direction. The depth of the face, typically very small, may be machined in a single pass or may be reached by machining at a smaller axial depth of cut and making multiple passes. For a finish operation, the cutting feed is calculated based on the desired surface roughness and the tool nose radius.

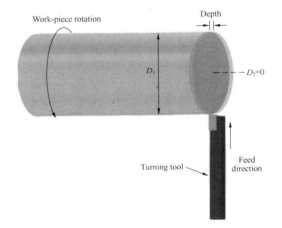

EXERCISE 7

This section is to test your ability to identify following operations.

| Bored contour   Facing to inner diameter   End milling |

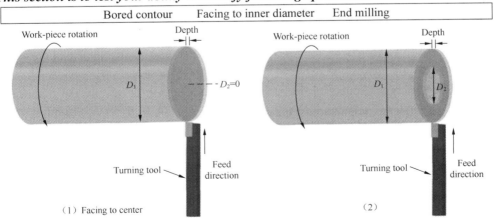

(1) Facing to center

(2)

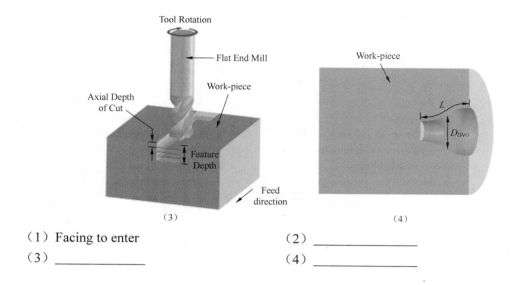

(1) Facing to enter　　　　　　　　(2) _____
(3) _____　　　　　　　　　　　　(4) _____

Part IV Broaden Your Horizon—Practical Activity

Facing Operations

Facing is the process of removing metal from the end of a work-piece to produce a flat surface.

Preparing for the Facing Cut

First, make sure the tumbler gear lever on the back of the headstock is in the neutral (middle) position so that the lead-screw does not rotate.

Clamp the work-piece tightly in the three-jaw chuck. To get the work properly centered, close the jaws until they just touch the surface of the work, then rotate the work-piece by hand in the jaws to seat it; then tighten the jaws (Figure 2-17).

Figure 2-17

Beginning the Facing Cut

Use the compound handwheel to advance the tip of the tool until it just touches the end of the work-piece. Use the cross feed crank to back off the tool until it is beyond the diameter of the work-piece. Turn the lathe on and adjust the speed to a few hundred RPM — setting of the speed control knob. Now slowly advance the cross feed hand-wheel to move the tool towards the work-piece. When the tool touches the work-piece it should start to remove metal from the end (Figure 2-18). Continue advancing the tool until it reaches the center of the work-piece and then crank the tool back in the opposite direction (towards you) until it is back past the edge of the work-piece (Figure 2-19).

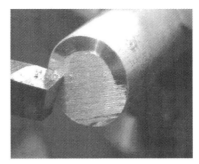

Figure 2-18   Figure 2-19

Since we started with the tool just touching the end of the work-piece, you probably removed very little metal on this pass. This is a good idea until you get used to how aggressively you can remove metal without stalling the lathe.

The Roughing Cut

Use the compound crank to advance the tool towards the chuck about 0.010". If the compound is set at a 90 degrees to the cross slide (which is how I usually set mine) then each division you turn the crank will advance the tool 0.001 (one-thousandth of an inch) toward the chuck.

If the compound is set at some other angle, say 30 degrees, to the cross slide, then it will advance the tool less than 0.001 for each division. The exact amount is determined by the trigonometric sine of the angle. Here's a picture of the first pass of a facing operation (Figure 2-20).

Figure 2-20

Cutting on the Return Pass

If you crank the tool back towards you after it reaches the center of the work-piece you will notice that it removes a small amount of metal on the return pass. This is because the surface is not perfectly smooth and it is removing metal from the high spots. If you need to remove a lot of metal, to reduce the work-piece to a specific length, for example, you can take advantage of this return cut to remove more metal on each pass by advancing the tool a small ways into the work-piece on the return pass. Since the tool must plunge into the face of the work-piece, this works best with a fairly sharp pointed tool.

The Finishing Cut

Depending on how rough the end of the work-piece was to begin with and how large the diameter is, you may need to make three or more passes to get a nice smooth finish across the face. These initial passes are called roughing passes and remove a relatively large amount of metal.

When you get the face pretty smooth you can make a final finishing cut to remove just 0.001 to 0.003" of metal and get a nice smooth surface. Figure 2-21 shows removing about 0.002" of metal at around 1 000 RPM. Figure 2-22 shows the finished face of the work-piece.

Figure 2-21

Figure 2-22

Figure 2-23 shows what happens if the tip of your cutting tool is below the center line of the lathe — a little nub is left at the center of the work-piece. The same thing happens if the tool is too high but the nub will have more of a cone shape in that case. If the tool is too low, place a suitable thickness of shim stock underneath the tool in the tool holder. If it's too high, grind the top down a few thous.

Filing the Edge

Facing operations leave a rather sharp edge on the end of the work-piece. It's a good idea to smooth this edge down with a file to give it a nice chamfer and to avoid cutting yourself on it. With the lathe running at fairly low speed, bring a smooth cut file up to the end of the work-piece at a 45 degree angle and apply a little pressure to the file. Figure 2-24: Right — left hand holding tang end of file. Figure 2-25: Wrong — left hand reaching over spinning chuck!

Figure 2-23

Figure 2-26 shows the finished surface and beveled edge. This is what a good facing cut should look like: smooth even surface with no raised bump in the center. Lay an accurate straight edge across the surface of the face and you should not be able to see any light under the edge. If you detect a slight convex shape, the carriage may be moving back away from the headstock during the cut.

Figure 2-24   Figure 2-25   Figure 2-26

Unit 3
Machining Operations and Turning Machines

| Part I Technical and Practical Reading |

Passage A   Machining Operations

Conventional machining, one of the most important material removal methods, is a collection of material-working processes in which power-driven machine tools, such as lathes, milling machines, and drill presses are used with a sharp cutting tool to mechanically cut the material to achieve the desired geometry. Machining is a part of the manufacture of almost all metal products. It is not uncommon for other materials to be machined. A person who specializes in machining is called a machinist. Machining is also a hobby. A room, building, or company where machining is done is called a machine shop. Much of modern day machining is controlled by computers using Computer Numerical Controlled (CNC) machining.

Machining Operations

The three principal machining processes are classified as turning, drilling and milling. Other operations falling into miscellaneous categories include shaping, planning, boring, broaching and sawing (Figure 3-1).

Figure 3-1   Machining operation

Turning operations are operations that rotate the work-piece as the primary method of moving metal against the cutting tool. Lathes are the principal machine tool used in turning.

Milling operations are operations in which the cutting tool rotates to bring cutting edges to bear against the work-piece. Milling machines are the principal machine tool used in milling.

Drilling operations are operations in which holes are produced or refined by bringing a rotating cutter with cutting edges at the lower extremity into contact with the work-piece. Drilling operations are done primarily in drill presses but not uncommonly on lathes or mills.

More recent advanced machining techniques include electrical discharge machining (EDM), electro-chemical erosion, laser, or water jet cutting to shape metal work-pieces. Machining requires attention to many details for a work-piece to meet the specifications set out in the engineering drawings or blueprints.

Types of Machining Operation

There are many kinds of machining operations, each of which is capable of generating a certain part geometry and surface texture.

In turning, a cutting tool with a single cutting edge is used to remove material from a rotating work-piece to generate a cylindrical shape. The speed motion in turning is provided by the rotating work-part, and the feed motion is achieved by the cutting tool moving slowly in a direction parallel to the axis of rotation of the work-piece (Figure 3-2).

Drilling is used to create a round hole. It is accomplished by a rotating tool that typically has two cutting edges. The tool is fed in a direction parallel to its axis of rotation into the work-part to form the round hole (Figure 3-3).

In boring, the tool is used to enlarge an already available hole. It is a fine finishing operation used in the final stages of product manufacture.

In milling, a rotating tool with multiple cutting edges is moved slowly relative to the material to generate a plane or straight surface. The direction of the feed motion is perpendicular to the tool's axis of rotation. The speed motion is provided by the rotating milling cutter (Figure 3-4).

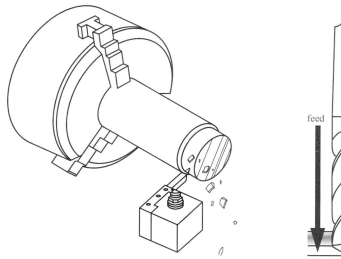

Figure 3-2 Turning

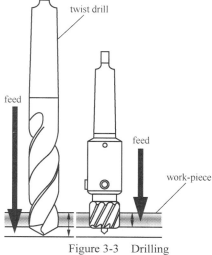

Figure 3-3 Drilling

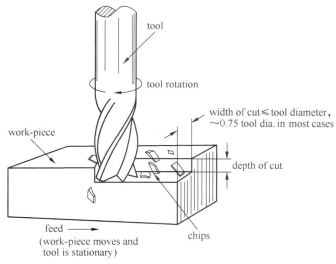

Figure 3-4 Milling

Overview of Machining Technology

Machining is not just one process; it is a group of processes. The common feature is the use of a cutting tool to form a chip that is removed from the work-part, called swarf. To perform the operation, relative motion is required between the tool and work. This relative motion is achieved in most machining operation by means of a primary motion, called cutting speed and a secondary motion called feed. The shape of the tool and its penetration into the work surface, combined with these motions, produce the desired shape of the resulting work surface (Figure 3-5).

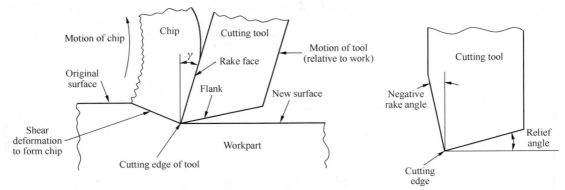

Figure 3-5　Basic machining processes

Notes

1. Conventional machining, one of the most important material removal methods, is a collection of material-working processes in which power-driven machine tools, such as lathes, milling machines, and drill presses are used with a sharp cutting tool to mechanically cut the material to achieve the desired geometry. 在这个长句中，主语是 Conventional machining，由 is 引导出合成谓语；one of the most…methods 是插入语；in which 引导定语从句，其先行词为 processes。在定语从句中，are used 是定语从句中的谓语动词。此句可译为：常规加工是集各种材料加工工艺于一体的最重要的切除材料的方法之一，在此过程中电动机床，如车床、铣床和钻床，利用锋利的刀具来进行机械加工以获得期望的几何形状。

2. Drilling operations are operations in which holes are produced or refined by bringing a rotating cutter with cutting edges at the lower extremity into contact with the work-piece. 在此句中，in which 引导定语从句，修饰先行词 operations。by 表示方法、方式。bing… into contact with… 意为"使……与……接触"。本句可译为：钻削是通过刀具旋转对工件进行钻孔或孔的精加工的一种操作，在钻头下面切入工件的顶端带有两个切削刃。

3. The shape of the tool and its penetration into the work surface, combined with these motions, produce the desired shape of the resulting work surface. 在此句中，主语是由 and 连接的并列主语，谓语动词为 produce。combined with…是分词作状语。本句意思为：刀具的形状、刀具对工件表面的切入加上它们之间的运动生成了最终的工件表面。

New Words

machining [məˈʃiːniŋ]　n. 加工
conventional [kənˈvenʃənl]　a. 常规的
removal [riˈmuːvəl]　n. 切削
power-driven [ˈpauəˈdrivn]　a. 电动的
mechanically [miˈkænikəli]　ad. 机械地
geometry [dʒiˈɔmitri]　n. 几何形状

Unit 3 Machining Operations and Turning Machines

manufacture [ˌmænjuˈfæktʃə] n. 制造
miscellaneous [misiˈleinjəs] a. 其他的
shape [ʃeip] v. 牛头刨削
plane [plein] v. 龙门刨削
bore [bɔ:] v. 镗孔
broach [brəutʃ] v. 拉削
edge [edʒ] n. 刀口
refine [riˈfain] v. 精加工
extremity [iksˈtremiti] n. 末端
discharge [disˈtʃɑ:dʒ] n. 放电
electro-chemical [iˌlektrəuˈkemikəl] a. 电化学的
erosion [iˈrəuʒən] n. 腐蚀
laser [ˈleizə] n. 激光
jet [dʒet] n. 喷射
specification [ˌspesifiˈkeiʃən] n. 技术规范
blueprint [ˈblu:ˌprint] n. 蓝图，设计图
generate [ˈtekstʃə] v. 生成
texture [ˈtekstʃə] n. 结构，纹理
cylindrical [siˈlindrik(ə)l] a. 圆柱的
enlarge [inˈlɑ:dʒ] v. 扩大
available [əˈveiləbl] a. 已有的
multiple [ˈmʌltipl] a. 多个的
perpendicular [ˌpə:pənˈdikjulə] a. 垂直的
cutter [ˈkʌtə] n. 刀具
swarf [swɔ:f] n. 金属切屑
secondary [ˈsekəndəri] a. 次要的
penetration [peniˈtreiʃən] n. 切入，穿透
resulting [riˈzʌltiŋ] a. 结果为，作为结果为

Phrases and Expressions

a collection of 很多，一批
drill press 钻床
specialize in 专门从事
fall into 分类为
more recent 最近
electrical discharge machining (EDM) 电火花加工
set out 做出
be parallel to 与……平行

axis of rotation　　旋转轴
fine finishing　　精加工
by means of　　通过
primary motion　　主运动
secondary motion　　次运动
be combined with　　被动与……结合

EXERCISE 1

The following is set of symbols denoting Working Safety. Choose the best symbol according to the information given below.

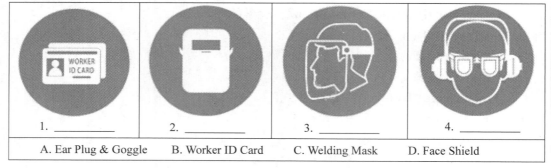

1. _____　　2. _____　　3. _____　　4. _____

A. Ear Plug & Goggle　　B. Worker ID Card　　C. Welding Mask　　D. Face Shield

EXERCISE 2

Translate the following phrases into Chinese or English.

1. conventional machining　_____
2. _____　铣床
3. numerical control　_____
4. _____　放电
5. chemical erosion　_____
6. _____　工程图
7. surface texture　_____
8. _____　进给运动

Passage B   Turning Machines

　　Turning machines, typically referred to as lathes, can be found in a variety of sizes and designs. While most lathes are horizontal turning machines, vertical machines are sometimes used, typically for large diameter work-pieces. Turning machines can also be classified by the type of control that is offered. A manual lathe requires the operator to control the motion of the cutting tool during the turning operation. Turning machines are also able to be computer controlled, in which case they are referred to as a computer numerical control (CNC) lathe. CNC lathes rotate the work-piece and move the cutting tool based on commands that are preprogrammed and offer very high precision. In this variety of turning machines, the main components that enable the work-piece to be rotated and the cutting tool to be fed into the work-piece remain the same. These components include the following (Figure 3-6).

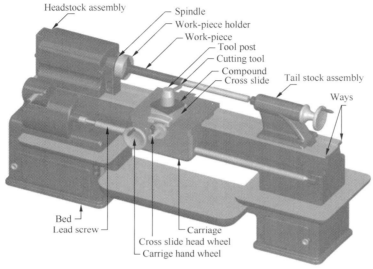

Figure 3-6　Manual Lathe

Bed

The bed of the turning machine is simply a large base that sits on the ground or a table and supports the other components of the machine (Figure 3-7).

Headstock Assembly

The headstock assembly is the front section of the machine that is attached to the bed. This assembly contains the motor and drive system which powers the spindle. The spindle supports and rotates the work-piece, which is secured in a work-piece holder or fixture, such as a chuck or collet (Figure 3-8).

Figure 3-7　Bed　　　　　　　　　　　　　Figure 3-8　Headstock assembly

Tailstock Assembly

The tailstock assembly is the rear section of the machine that is attached to the bed. The purpose of this assembly is to support the other end of the work-piece and allow it to rotate, as it's driven by the spindle. For some turning operations, the work-piece is not supported by the tailstock so that material can be removed from the end (Figure 3-9).

Carriage

The carriage is a platform that slides alongside the work-piece, allowing the cutting tool to cut away material as it moves. The carriage rests on tracks that lay on the bed, called "ways", and is advanced by a lead screw powered by a motor or hand wheel (Figure 3-10).

Figure 3-9   Tail stock assembly   Figure 3-10   Carriage

Cross Slide

The cross slide is attached to the top of the carriage and allows the tool to move towards or away from the work-piece, changing the depth of cut. As with the carriage, the cross slide is powered by a motor or hand wheel (Figure 3-11).

Figure 3-11   Cross slide head wheel

Unit 3 Machining Operations and Turning Machines

Compound

The compound is attached on top of the cross slide and supports the cutting tool. The cutting tool is secured in a tool post which is fixed to the compound. The compound can rotate to alter the angle of the cutting tool relative to the work-piece.

Turret

Some machines include a turret, which can hold multiple cutting tools and rotates the required tool into position to cut the work-piece. The turret also moves along the work-piece, feeding the cutting tool into the material. While most cutting tools are stationary in the turret, live tooling can also be used. Live tooling refers to powered tools, such as mills, drills, reamers, and taps, which rotate and cut the work-piece (Figure 3-12).

Figure 3-12 Turret

New Words

horizontal [ˌhɔriˈzɔntl]   a. 水平式的
vertical [ˈvəːtikəl]   a. 垂直式的
classify [ˈklæsifai]   v. 分类
manual [ˈmænjuəl]   a. 手动的
operator [ˈɔpəreitə]   n. 操作者
numerical [njuː(ː)ˈmerikəl]   a. 数字的
command [kəˈmɑːnd]   n. 指令
preprogrammed [priˈprəugræmd]   a. 预先编制好的
precision [priˈsiʒən]   n. 精度
feed [fiːdç]   v. 供料，提供
secure [siˈkjuə]   v. 固定
holder [ˈhəuldə]   n. 固定装置
fixture [ˈfikstʃə]   n. 夹具
collet [ˈkɔlit]   n. 棘爪
platform [ˈplætfɔːm]   n. 平台
track [træk]   n. 轨道
advance [ədˈvɑːns]   v. 向前运动
compound [ˈkɔmpaund]   n. 复合刀架
alter [ˈɔːltə]   v. 改变
angle [ˈæŋgl]   n. 角度
turret [ˈtʌrit]   n. 转塔

live [laiv]   a. 活动的

tooling ['tu:liŋ]   n. 工具，工装

Phrases and Expressions

be referred to as   被称为

a variety of   多种

computer numerical control (CNC)   计算机数控

headstock assembly   主轴箱

tailstock assembly   尾架

rest on   位于

hand wheel   手轮

cross slide   横刀架

depth of cut   切削深度

tool post   刀座

powered tool   电动工具

EXERCISE 3

Choose the best machining way according to the information given.

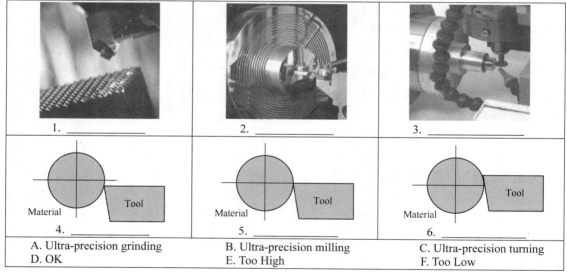

A. Ultra-precision grinding
D. OK
B. Ultra-precision milling
E. Too High
C. Ultra-precision turning
F. Too Low

EXERCISE 4

Abbreviations are very useful in practical work. Read them and then translate them into corresponding Chinese terms.

1. CMI	Computer Managed Instruction	_____
2. CRS	Cold-rolled Steel	_____
3. CS	Complete Self-protecting	_____

Unit 3 Machining Operations and Turning Machines

4. CO. NI.	Copper Nickel Alloy	_____	
5. DC	Direct Current	_____	
6. DC rel.	Direct Current Relay	_____	
7. DME	Distance Measuring Equipment	_____	
8. DP	Difference of Potential	_____	

Part II Glance at Conventional Machine Tool Structures

The following is the structure of a conventional machine tool.

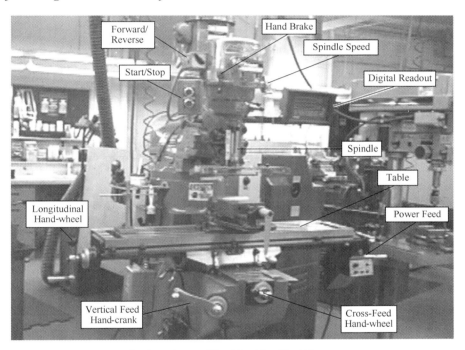

Explanations of the Machine Tool Terms

No.	Name	Explanation
1	Forward/Reverse	正转/反转
2	Start/Stop	启动/停止
3	Longitudinal Hand-wheel	纵向手轮
4	Vertical Feed Hand-crank	垂直进给手柄
5	Cross-Feed Hand-wheel	横向进给手轮
6	Power Feed	电动进给
7	Table	工作台
8	Spindle	主轴
9	Digital Readout	数码屏
10	Spindle Speed	主轴转速
11	Hand Brake	手动制动器

EXERCISE 5

The following is the conventional machine Tool. You are required to choose the suitable words or phrases given below.

Double Boring Machine   Vertical Diamond Fine Boring Machine   Deep Hole Boring Machine   CNC Plano-Boring and Milling Machine   Horizontal Milling and Boring Machine

Boring Machine（镗床）

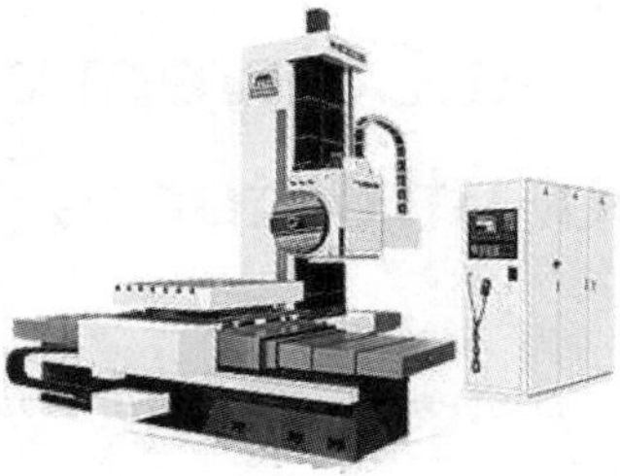

1. _____ （卧式铣镗床）

2. _____ （立式金刚镗床）

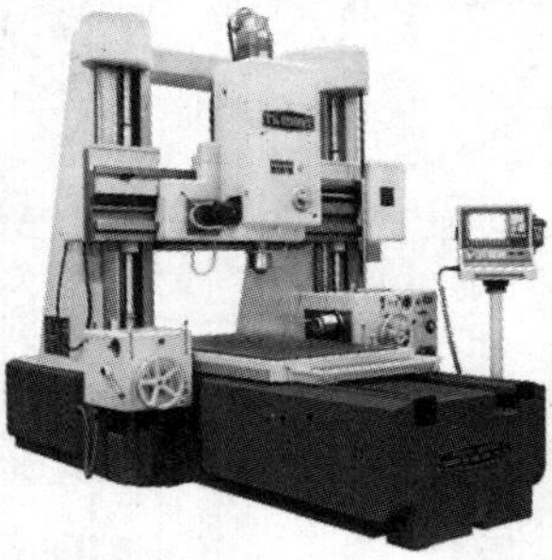

3. _____ （双坐标镗床）

4. _____ （深孔镗床）

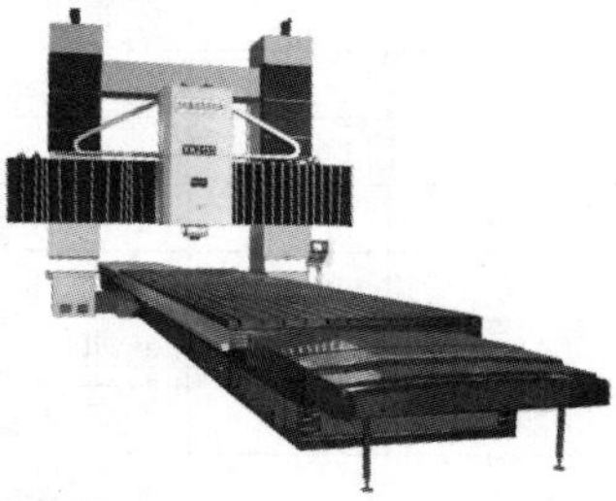

5. _____ （数控龙门镗铣床）

Unit 3 Machining Operations and Turning Machines

| Part III   Simulated Writing |

Section A   Match Your Skill

The following is a lathe accessory, and you can understand the names of Machine Vice.

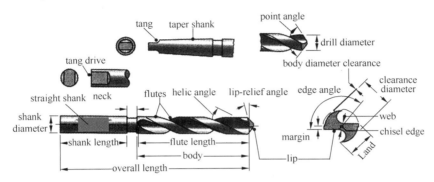

Explanations of the Terms

No.	Name	Explanation
1	twist bits	麻花钻
2	tang	扁尾
3	taper shank	锥柄
4	point angle	顶尖角
5	tang drive	扁尾驱动
6	flutes	韧带
7	helic angle	螺旋角
8	lip-relief angle	刃倾角
9	edge angle	横刀斜角
10	drill diameter	钻头直径
11	body diameter clearance	刀体直径间隙
12	clearance diameter	间隙直径
13	straight shank	直柄
14	neck	颈部
15	shank diameter	刀柄直径
16	shank length	刀柄长度
17	flute length	韧带长度
18	body	刀体
19	overall length	总长
20	margin	刃边
21	lip	刀刃
22	web	钻心
23	chisel edge	横刃

EXERCISE 6

Match the words or phrases on the left with their meanings on the right.

Solid Drill Head Structure

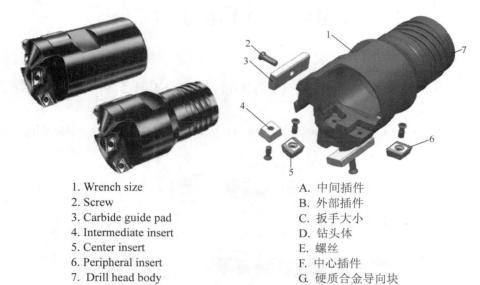

1. Wrench size
2. Screw
3. Carbide guide pad
4. Intermediate insert
5. Center insert
6. Peripheral insert
7. Drill head body

A. 中间插件
B. 外部插件
C. 扳手大小
D. 钻头体
E. 螺丝
F. 中心插件
G. 硬质合金导向块

Section B Have a Try

This section will help you to understand several forms of machining.

External Operations

Grooving —A turning operation in which a single-point tool moves radially, into the side of the work-piece, cutting a groove equal in width to the cutting tool. If the desired groove width is larger than the tool width, multiple adjacent grooves will be cut. A profiling cut can be performed to smooth the surface of multiple grooves. Special form tools can also be used to create grooves of varying geometries.

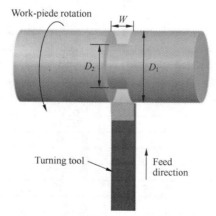

EXERCISE 7

This section is to test your ability to identifying different operations.

| Form tool | Peripheral cut | End mill |

Unit 3 Machining Operations and Turning Machines

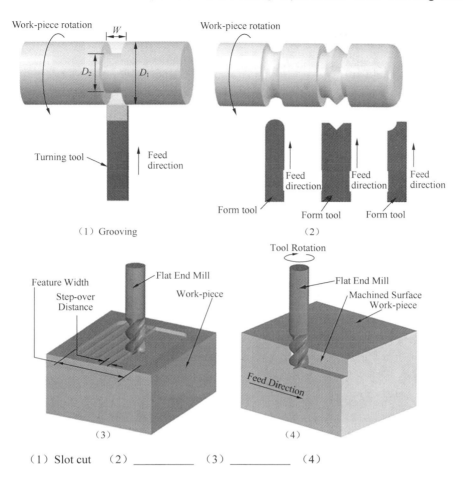

(1) Slot cut　(2) _____　(3) _____　(4)

Part IV   Broaden Your Horizon—Practical Activity

Turning Operations

Turning is the removal of metal from the outer diameter of a rotating cylindrical work-piece. Turning is used to reduce the diameter of the work-piece, usually to a specified dimension, and to produce a smooth finish on the metal.

Chucking the Work-piece

We will be working with a piece of 3/4" diameter 6061 aluminum about 2 inches long. We can safely turn it in the three jaw chuck without supporting the free end of the work.

Insert the work-piece in the three-jaw chuck and tighten down the jaws until they just start to grip the work-piece. Rotate the work-piece to ensure that it is seated evenly and to dislodge any chips. You want the work-piece to be as parallel as possible with the center line of the lathe. Imagine an exaggerated example where the work-piece is skewed at an angle in the chuck and you can easily visualize why this is important. Tighten the chuck using each of the three chuck key positions to ensure a tight and even grip (Figure 3-13).

Adjusting the Tool Bit

Choose a tool bit with a slightly rounded tip. This type of tool should produce a nice smooth finish. Adjust the angle of the tool-holder so the tool is approximately perpendicular to the side of the work-piece. Because the front edge of the tool is ground at an angle, the left side of the tip should engage the work, but not the entire front edge of the tool. The angle of the compound is not critical. I usually keep mine at 90 degrees so that the compound dial advances the work (Figure 3-14).

Figure 3-13

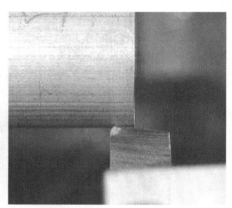
Figure 3-14

Cutting Speeds

You must consider the rotational speed of the work-piece and the movement of the tool relative to the work-piece. Working with the 7×10 for hobby purposes, you will quickly develop a feel for how fast you should go.

Until you get a feel for the proper speeds, start with relatively low speeds and work up to faster speed. Most cutting operations on the 7×10 will be done at speeds of a few hundred RPM —with the speed control set below the 12 o'clock position and with the HI/LO gear in the LO range.

Turning with Hand Feed

As always, wear safety glasses and keep your face well away from the work since this operation will throw off hot chips and/or sharp spirals of metal.

Now advance the cross slide crank about 10 divisions or 0.010". Turn the carriage hand-wheel counterclockwise to slowly move the carriage towards the headstock. As the tool starts to cut into the metal, maintain a steady cranking motion to get a nice even cut. It's difficult to get a smooth and even cut turning by hand (Figure 3-15).

Continue advancing the tool towards the headstock until it is about 1/4" away from the chuck jaws. Obviously you want to be careful not to let the tool touch the chuck jaws (Figure 3-16)!

Unit 3 Machining Operations and Turning Machines

Figure 3-15

Figure 3-16

Turning with Power Feed

One of the great features of the 7×10 is that it has a power lead-screw driven by an adjustable gear train. The lead-screw can be engaged to move the carriage under power for turning and threading operations.

To change the lever setting, you must pull back on the knurled sleeve with considerable force. With the sleeve pulled back you can move the lever up and down to engage its locking pin in one of three positions. In the upper position the lead-screw rotates to move the carriage towards the headstock and in the lower position the lead-screw moves the carriage away from the headstock (Figure 3-17).

In the down position, the half-nut lever engages two halves of a split nut around the lead-screw. Make sure the half-nut lever is in the disengaged (up) position. Turn the motor on. The lead-screw should now be rotating counterclockwise (Figure 3-18). When the lead-screw is engaged the gear train makes kind of an annoying noise, but you'll get used to it. Lubricating the gear train with white lithium grease will cut down some on the noise.

Figure 3-17

Figure 3-18

Just as in facing, you normally will make one or more relatively deep (0.010~0.030) roughing cuts followed by one or more shallow (0.001~0.002) finishing cuts. Of course you have to plan these cuts so that the final finishing cut brings the work-piece to exactly the desired diameter (Figure 3-19).

When cutting under power, you must be very careful not to run the tool into the chuck.

Measuring the Diameter

It is important to recognize that, in a turning operation, each cutting pass removes twice the amount of metal indicated by the cross slide feed divisions. Therefore, when advancing the cross slide by 0.010", the diameter is reduced by 0.020".

The diameter of the work-piece is determined by a caliper or micrometer. Micrometers are more accurate, but less versatile. You will need a machinist's caliper capable of measuring down to 0.001". Vernier calipers do not have a dial and require you to interpolate on an engraved scale (Figure 3-20).

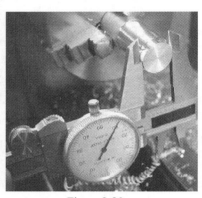

Figure 3-19   Figure 3-20

Turning a Shoulder

A shoulder is a point at which the diameter of the work-piece changes with no taper from one diameter to the other.

We will make a shoulder on our work-piece by reducing the diameter of the end of the work-piece for a distance of about 1/2" (Figure 3-21). Advance the cross slide about 0.020 and use power feed to turn down about a 1/2" length on the end of the work-piece. Repeat this a few more times until you have reduced the diameter of the end section to about 1/2" (Figure 3-22).

Since the tip of the tool is rounded, the inner edge of the shoulder takes on a rounded profile.

Now we will use this pointed tool to make a square finishing cut into the corner of the shoulder. Since this is such a short distance, we will use hand feed, not power feed. You can use hand feed with the lead-screw turning — just don't engage the half-nut (Figure 3-23).

To get a nice square face on the shoulder you will need to make a facing cut. This works best if you have made a carriage lock on your lathe. Lock the carriage and clean up the face of the shoulder until it is square. If you use the sharp pointed tool you will need to use fairly high RPM, say 1500, and advance the tool slowly or you will get little grooves from the pointed tip instead of a nice smooth finish.

Finally, you may want to use a file as described in the facing section to make a nice beveled edge on outside edge of the shoulder and on the end of the work-piece (Figure 3-24).

Figure 3-21

Figure 3-22

Figure 3-23

Figure 3-24

Unit 4
Hydraulic Machinery and Forging Equipment

| Part I   Technical and Practical Reading |

Passage A   Hydraulic Machinery

Hydraulic machinery is machines and tools which use fluid power to do work. In this type of machine, high-pressure liquid — called hydraulic fluid — is transmitted throughout the machine to various hydraulic motors and hydraulic cylinders.

Hydraulic machinery is operated by the use of hydraulics, where a liquid is the powering medium. Pneumatics, on the other side, is based on the use of gas as the medium for power transmission, generation and control.

Force and Torque Multiplication

A fundamental feature of hydraulic systems is the ability to apply force or torque multiplication in an easy way without the need of mechanical gears or levers, either by altering the effective areas in two connected cylinders or the effective displacement between a pump and motor (Figure 4-1, Figure 4-2). Both these examples are usually referred to as a hydraulic transmission or hydrostatic transmission involving a certain hydraulic "gear ratio".

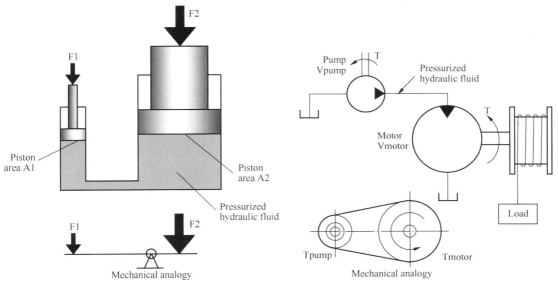

Figure 4-1 Force increase with hydraulics

Figure 4-2 Torque increase with hydraulics

Hydraulic Circuits

Open center circuits use pumps which supply a continuous flow. The flow is returned to tank through the control valve's open center; that is, when the control valve is centered, it provides an open return path to tank and the fluid is not pumped to a high pressure (Figure 4-3).

Closed center circuits supply full pressure to the control valves, whether any valves are actuated or not. The pumps vary their flow rate, pumping very little hydraulic fluid until the operator actuates a valve. The valve's spool therefore doesn't need an open center return path to tank. Multiple valves can be connected in a parallel arrangement and system pressure is equal for all valves (Figure 4-4).

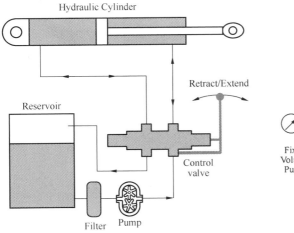

Figure 4-3 Open center circuits

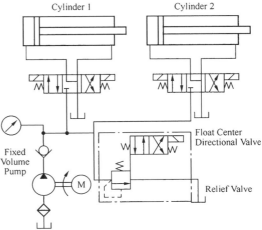

Figure 4-4 Closed center circuits

Open and closed circuits

Open-loop: Pump-inlet and motor-return (via the directional valve) are connected to the hydraulic tank. The term loop applies to feedback; the more correct term is open versus closed "circuit" (Figure 4-5).

Closed-loop: Motor-return is connected directly to the pump-inlet. To keep up pressure on the low pressure side, the circuit has a charge pump that supplies cooled and filtered oil to the low pressure side. Closed-loop circuits are generally used for hydrostatic transmissions in mobile applications (Figure 4-6).

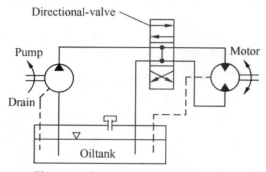

Figure 4-5 Open-loop hydraulic circuits

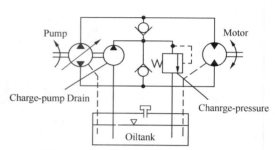

Figure 4-6 Closed-loop hydraulic circuits

Advantages: No directional valve and better response, the circuit can work with higher pressure. The pump swivel angle covers both positive and negative flow direction.

Disadvantages: The pump cannot be utilized for any other hydraulic function in an easy way and cooling can be a problem due to limited exchange of oil flow.

Hydraulic Pump

Hydraulic pumps supply fluid to the components in the system. Pressure in the system develops in reaction to the load. Hence, a pump rated for 5000 psi is capable of maintaining flow against a load of 5000 psi.

Pumps have a power density about ten times greater than an electric motor (by volume). They are powered by an electric motor or an engine, connected through gears, belts, or a flexible elastomeric coupling to reduce vibration (Figure 4-7).

Control Valve

Directional control valves are usually designed to be stackable, with one valve for each hydraulic cylinder, and one fluid input supplying all the valves in the stack (Figure 4-8).

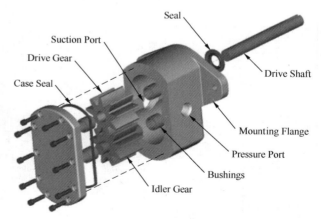

Figure 4-7 Hydraulic pump

Unit 4 Hydraulic Machinery and Forging Equipment

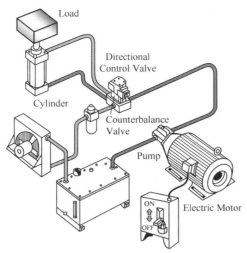

Figure 4-8 Control valve

Directional control valves route the fluid to the desired actuator. They usually consist of a spool inside a cast iron or steel housing. The spool slides to different positions in the housing, intersecting grooves and channels route the fluid based on the spool's position.

Notes

1. A fundamental feature of hydraulic systems is the ability to apply force or torque multiplication in an easy way without the need of mechanical gears or levers, either by altering the effective areas in two connected cylinders or the effective displacement between a pump and motor. 此句中的"either…or…"意为"……或……"，连接句子中的两个并列成分，即 the effective areas 和 the effective displacement。"by"表示手段、方法。本句可译为：液压系统的一个基本特征是能够在不需要机械齿轮或杠杆的情况下，轻松地使力或力矩倍增。这是通过改变两个相连液压缸间的有效面积或改变泵和电机之间的有效排量来实现的。

2. To keep up pressure on the low pressure side, the circuits have a charge pump that supplies cooled and filtered oil to the low pressure side. "to keep up pressure on the low pressure side" 不定式作目的状语。"that"引导定语从句，先行词为"a charge pump"。本句可译为：为了保持低压侧的压力，回路有个补油泵，用来提供经过冷却和过滤的液压油给低压侧。

3. The spool slides to different positions in the housing, intersecting grooves and channels route the fluid based on the spool's position. 这是一个由 and 连接的并列句。两个简单句的主语分别为 the spool 和 channels，谓语分别为 slides 和 route。整句可译为：阀芯滑移到阀体中的不同位置，阀芯上的环槽与阀体上的通道就根据这个阀芯的位置来分配液压油的流向。

New Words

hydraulic [haiˈdrɔːlik]　　adj. 液压的，水压的
transmit [trænzˈmit]　　vt. 传输，转送

pneumatics [njuːˈmætiks]   n. 气压工程，气体力学
machinery [məˈʃiːnəri]   n.（总称）机器，机械
forge [fɔːdʒ]   v. 锻造
fluid [ˈfluː(ː)id]   n. 流体
hydraulics [ˈhaiˈdrɔːliks]   n. 液压技术，水力学
transmission [trænzˈmiʃən]   n. 传输，传送
medium [ˈmiːdiəm]   n. 介质
torque [tɔːk]   v. 施以转动力
             n. 转力矩
multiplication [ˌmʌltipliˈkeiʃən]   n. 增加，倍增
lever [ˈliːvə, ˈlevə]   n. 杠杆
             v. 撬起
displacement [disˈpleismənt]   n. 排水，位移量
hydrostatic [ˌhaidrəuˈstætik]   a. 静水力学的
analogy [əˈnælədʒi]   n. 模拟
circuit [ˈsəːkit]   n. 回路，电路
tank [tæŋk]   n. 水槽
actuate [ˈæktjueit]   v. 开动（机器），使运转
spool [spuːl]   n. 线轴
parallel [ˈpærəlel]   a. 并联的，平行的
arrangement [əˈreindʒmənt]   n. 排列
reservoir [ˈrezəvwɑː]   n. 流体箱
filter [ˈfiltə]   n. 过滤器   v. 过滤
retract [riˈtrækt]   v. 拉回
extend [iksˈtend]   v. 拉出
inlet [ˈinlet]   n. 入口
drain [drein]   n. 泄油
swivel [ˈswivl]   v. 使旋转，使回旋
             n. 转轴
bushing [ˈbuʃiŋ]   n. 隔套
port [pɔːt]   n. 孔
seal [siːl]   n. 密封套
component [kəmˈpəunənt]   n. 元件，组件
belt [belt]   n. 传送带
coupling [ˈkʌpliŋ]   n. 联轴器
vibration [vaiˈbreiʃən]   n. 振动
stackable [ˈstækəbl]   a. 叠加式的
counterbalance [ˌkauntəˈbæləns]   n. 平衡，配重
actuator [ˈæktjueitə]   n. 执行元件，执行装置

Unit 4 Hydraulic Machinery and Forging Equipment

intersect [ˌintəˈsekt] v. 相交于
housing [ˈhauziŋ] n. 外壳，外罩
groove [gruːv] n. 凹槽，环槽
route [ruːt] n. 流道
　　　　　　v. 按某路线传递

Phrases and Expressions

hydraulic transmission　液压传动
hydrostatic transmission　液压静力传动
pressurized hydraulic fluid　压力流体
open center circuit　开式回路
closed center circuit　闭式回路
control valve　控制阀
hydraulic cylinder　液压缸
fixed volume pump　定量泵
float center directional valve　Y 型中位机能电磁换向阀
relief valve　溢流阀，安全阀
directional valve　换向阀
hydraulic tank　油箱
charge pump　补油泵
oil tank　油箱
hydraulic pump　液压泵
case seal　密封垫
drive gear　主动齿轮
suction port　进油孔
idler gear　从动齿轮
pressure port　吸油孔
mounting flange　安装法兰
drive shaft　驱动轴
pounds per square inch (psi)　磅/平方英寸
power density　功率密度
counterbalance valve　弹簧复位阀
cast iron　铸铁

EXERCISE 1

The following is set of symbols denoting Working Safety. Choose the best symbol according to the information given below.

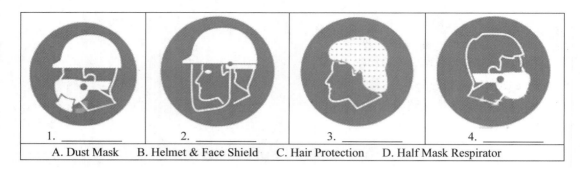

1. _____ 2. _____ 3. _____ 4. _____

A. Dust Mask   B. Helmet & Face Shield   C. Hair Protection   D. Half Mask Respirator

EXERCISE 2

Translate the following phrases into Chinese or English.

1. hydraulic machinery   _____
2. _____   动力传输
3. parallel arrangement   _____
4. _____   补油泵
5. swivel angle   _____
6. _____   液压泵
7. elastomeric coupling   _____
8. _____   功率密度

Passage B　Forging Equipment

Forging is the term for shaping metal by using localized compressive forces. Cold forging is done at room temperature or near room temperature. Hot forging is done at a high temperature, which makes metal easier to shape and less likely to fracture. Warm forging is done at intermediate temperature between room temperature and hot forging temperature. Forged parts can range in weight from less than a kilogram to 170 metric tons.

Hot Forging

Hot forging is defined as working a metal above its recrystallization temperature. The main advantage of hot forging is that as the metal is deformed the strain-hardening effects are negated by the recrystallization process (Figure 4-9).

Warm Forging

Warm forging has a number of cost-saving advantages which underscore its increasing use as a manufacturing method. The temperature range for the warm forging of steel runs from about 800 to 1800 degrees Fahrenheit. However, the narrower range of from 1000 to 1330 degrees Fahrenheit is emerging as the range of perhaps the greatest commercial potential for warm forging. Compared with cold forging, warm forging has the potential advantages of: reduced tooling loads, reduced press loads, increased steel ductility, elimination of need to anneal prior to forging, and favorable as-forged properties that can eliminate heat treatment (Figure 4-10).

Figure 4-9   Hot Forging

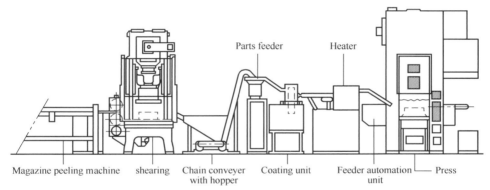

Figure 4-10   Warm Forging System

Figure 4-11   Cold Forging

Cold Forging

Cold forging is defined as working a metal below its recrystallization temperature, but usually around room temperature. If the temperature is above 0.3 times the melting temperature (on an absolute scale) then it qualifies as warm forging (Figure 4-11).

Process

Forging is a metal forming process used to produce large quantities of identical parts and to improve the mechanical properties of the metal being forged. The design of forged parts is limited when undercuts or cored sections are required. All cavities must be comparatively straight and largest at the mouth, so that the forging die may be withdrawn. The products of forging may be tiny or massive and can be made of steel (automobile axles), brass (water valves), tungsten (rocket nozzles), aluminum (aircraft structural members), or any other metal. Common forging processes include: roll forging, swaging, cogging, open-die forging, impression-die forging, press forging, automatic hot forging and upsetting.

Types of Forging

Forging is divided into three main methods: hammer, press, and rolled types.

1. Hammer Forging: Preferred method for individual forgings. The shaping of a metal, or other materials, by an instantaneous application of pressure to a relatively small area. A hammer or ram, delivering intermittent blows to the section to be forged, applies this pressure. The hammer is dropped from its maximum height, usually raised by steam or air pressure. Hammer forging can produce a wide variety of shapes and sizes and, if sufficiently reduced, can create a high degree of grain refinement at the same time. The disadvantage of this process is that finish machining is often required, as close dimensional tolerances cannot be obtained (Figure 4-12).

Figure 4-12   Hammer Forging

2. Press Forging: This process is similar to kneading, where a slow continuous pressure is applied to the area to be forged. The pressure will extend deep into the material and can be completed either cold or hot. A cold press forging is used on a thin, annealed material, and a hot press forging is done on large work such as armor plating, locomotives and heavy machinery. Press Forging is more economical than hammer forging (except when dealing with low production numbers), and closer tolerances can be obtained. A greater proportion of the work done is transmitted to the work-piece, differing from that of the hammer forging operation, where much of the work is absorbed by the machine and foundation (Figure 4-13).

3. Die Forging: Open and closed die operations can be used in forging. In open-die

Figure 4-13   Press Forging

forging the dies are either flat or rounded. Large forgings can be formed by successive applications of force on different parts of the material. Hydraulic presses and forging machines are both employed in closed die forging. In closed-die forging the metal is trapped in recessed impressions, which are machined into the top and bottom dies. As the dies press together, the material is forced to fill the impressions. Flash, or excess metal, is squeezed out between the dies. Closed-die forging can produce parts with more complex shapes than open-die forging. Die forging is the best method, as far as tolerances that can be met, and also results in a finished part that is completely filled out and is produced with the least amount of flashing. The final shape and the improvement in metallurgical properties are dependent on the skill of the operator (Figure 4-14).

 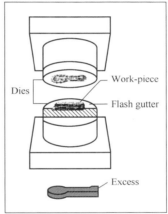

Figure 4-14　Die Forging

New Words

localized ['ləukəlaid]　a. 定位的
compressive [kəm'presiv]　a. 压缩的
fracture ['fræktʃə]　v. 断裂
recrystallization [riː'kristəlaizeiʃən]　n. 再结晶
deform [diː'fɔːm]　v. 变形
Fahrenheit ['færənhait]　n. 华氏度
ductility [dʌk'tiliti]　n. 延展性
anneal [ə'niːl]　v. 退火
strain-hardening [strein'hɑːdəniŋ]　a. 应变硬化的
negate [ni'geit]　v. 消除
press [pres]　n. 冲床
magazine [,mægə'ziːn]　n. 机台

shearing [ˈʃiəriŋ]　n. 切断加工
conveyer [kənˈveiə]　n. 输送机
hopper [ˈhɔpə]　n. 装料斗
coat [kəut]　v. 涂料
heater [ˈhi:tə]　n. 加热机
undercut [ˈʌndəkʌt]　n. 侧分型
cored [kɔ:d]　a. 芯型的
mouth [mauθ]　n. 浇口
withdraw [wiðˈdrɔ:]　v. 取出
nozzle [ˈnɔzl]　n. 喷管
member [ˈmembə]　n. 构件
swaging [ˈsweidʒiŋ]　n. 型锻
cogging [ˈkɔgiŋ]　n. 钝齿啮合
upsetting [ʌpˈsetiŋ]　n. 镦粗加工
blow [bləu]　n. 锤打
tolerance [ˈtɔlərəns]　n. 公差
knead [ni:d]　v. 捏制
foundation [faunˈdeiʃən]　n. 基座
recessed [riˈsesd]　a. 凹陷的
flash [flæʃ]　n. 飞边、溢料
squeeze [skwi:z]　v. 挤压
metallurgical [ˌmetəˈlə:dʒikəl]　a. 冶金的

Phrases and Expressions

compressive force　挤压力
cold forging　冷锻
hot forging　热锻
warm forging　温锻
prior to　在……之前
heat treatment　热处理
absolute scale　绝对温标
peeling machine　装料机
parts feeder　拾取定向料斗
coating unit　涂料机
feeder automation unit　自动进料机
melting temperature　熔化温度，熔点
mechanical property　机械特性
roll forging　辊锻

Unit 4 Hydraulic Machinery and Forging Equipment

open-die forging　开式模锻
impression-die forging　飞边模锻
press forging　压锻
forging die　锻模
hammer forging　锤锻
air pressure　气压
grain refinement　晶粒细化
dimensional tolerance　尺寸公差
finish machining　精加工
heavy machinery　重型机械
die forging　模锻
hydraulic press　液压机
forging machine　锻造机
closed-die forging　闭式模锻
top die　上模
bottom die　底模

EXERCISE 3

Match the operation stag with the picture given.

Isothermal Multistep Forging Process

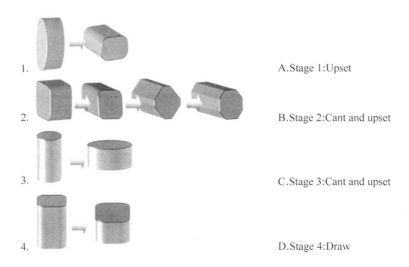

1.　　　　　　　　　　　A. Stage 1: Upset

2.　　　　　　　　　　　B. Stage 2: Cant and upset

3.　　　　　　　　　　　C. Stage 3: Cant and upset

4.　　　　　　　　　　　D. Stage 4: Draw

EXERCISE 4

Abbreviations are very useful in practical work. Read them and then translate them into corresponding Chinese terms.

1. DR	Differential Relay		
2. ED	Electron Device		
3. GOR	Gas Oil Ratio		
4. HP	High Power		
5. H.T.	Heat Treatment		
6. HSS	High Speed Steel		
7. IG	Involute Gear		
8. I/O	Input/Output		

| Part II Glance at Conventional Machine Tool Structures |

The following is a the structure of a regenerative hydraulic control.

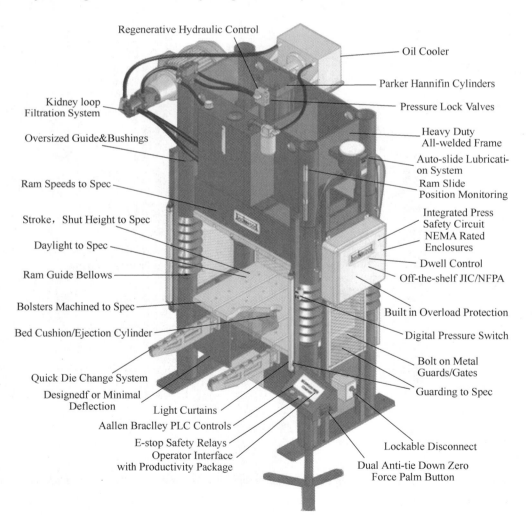

Unit 4 Hydraulic Machinery and Forging Equipment

Explanations of the Regenerative Hydraulic Control Terms

No.	Name	Explanation
1	Regenerative Hydraulic Control	回馈式液压控制
2	Kidney Loop Filtration System	肾形环过滤系统
3	Oversized Guide & Bushings	大型导杆和导套
4	Ram Speeds to Spec.	专业滑块
5	Stroke, Shut Height to Spec.	上滑块的最大行程
6	Daylight to Spec.	规格孔距
7	Ram Guide Bellows	波纹式防护罩
8	Bolsters Machined to Spec.	标准定位板
9	Bed Cushion/Ejection Cylinder	排屑槽
10	Quick Die Change System	床身减震/缓冲缸
11	Designed for Minimal Deflection	偏心凸轮
12	Light Curtains	灯罩
13	Allen Bradley PLC Controls	Allen Bradley 型可编程控制器
14	E-stop Safety Relays	急停按钮
15	Operator Interface with Productivity Package	操纵台
16	Dual Anti-Tie Down Zero Force Palm Button	双反常引力死限位按钮
17	Lockable Disconnect	行程限位
18	Guarding to Spec.	规格防护网
19	Bolt on Metal Guards /Gates	防护网螺栓
20	Digital Pressure Switch	压力传感器
21	Built in Overload Protection	内置过载保护电路
22	Off-the-shelf JIC/NFPA	JIC/NFPA 现成保护电路
23	Dwell Control	过压保护
24	NEMA Rated Enclosures	NEMA 型标准护罩
25	Integrated Press Safety Circuit	集成压力保护电路
26	Ram Slide Position Monitoring	位置传感器
27	Auto-slide Lubrication System	导轨润滑系统
28	Heavy Duty All-welded Frame	焊接结构件
29	Pressure Lock Valves	压力控制阀
30	Parker Hannifin Cylinders	Parker Hannifin 型油缸
31	Oil Cooler	散热器

EXERCISE 5

The following is a forging machine. You are required to choose the suitable words or phrases given below.

Flywheel Clutch and Brake   Eccentric Shaft   Connection   Press Bed
Slide (Ram) Shut Height Adjustment   Bolster   Counterbalance Cylinder
Saddle Bushing   Opposed Helical Gears at Each Side of Press

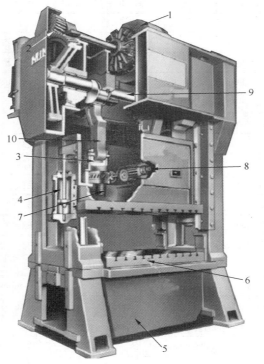

1. _____ 2. _____ 3. _____ 4. _____
5. _____ 6. _____ 7. _____ 8. _____
9. _____ 10. _____

Part III Simulated Writing

Section A Match Your Skill

The following is a lathe accessory, and you can understand the name of Bench Grinder.

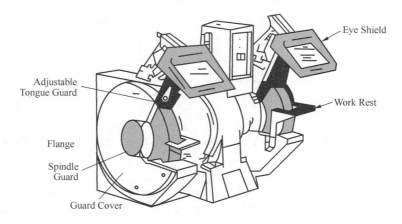

Unit 4　Hydraulic Machinery and Forging Equipment

Explanations of Terms

No.	Name	Explanation
1	Adjustable Tongue Guard	可调整的舌形护罩
2	Flange	法兰
3	Spindle Guard	主轴护罩
4	Eye Shield	护眼罩
5	Work Rest	工件架
6	Guard Cover	防护板

EXERCISE 6

Match the words or phrases on the left with their meanings on the right.

Bench Grinder

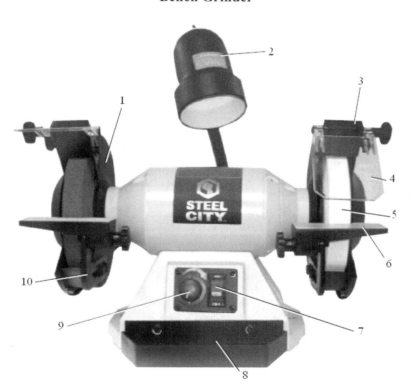

1. Inner Wheel Guard　　　　A. 阻火器
2. Flexible Work Light　　　　B. 砂轮
3. Spark Arrestor　　　　　　C. 内砂轮罩
4. Eye Shield　　　　　　　　D. 通/断开关
5. Grinding Wheel　　　　　　E. 灵活工作灯
6. Tool Rest　　　　　　　　　F. 变速旋钮
7. On/Off Switch　　　　　　　G. 护眼罩
8. Quench Tray　　　　　　　H. 可调节的刀架支撑器
9. Variable Speed Knob　　　　I. 刀架
10. Adjustable Tool Rest Support　J. 淬火盘

Section B Have a Try

This section will help you to understand several forms of machining.

External Operations

Cut-off — A turning operation, also known as parting, in which a single-point cut-off tool moves radially, into the side of the work-piece, and continues until the center or inner diameter of the work-piece is reached, thus parting or cutting off a section of the work-piece. A part catcher is often used to catch the removed part.

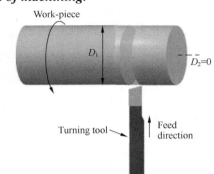

EXERCISE 7

This section is to test your ability to identifying following features of different operations.

End milling	Cut-off to inner diameter	Turning

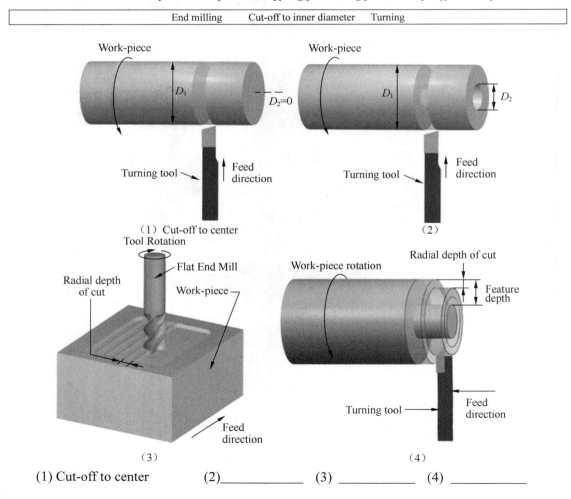

(1) Cut-off to center   (2)_____   (3) _____   (4) _____

| Part IV   Broaden Your Horizon—
Practical —Activity |

Forging

Other names/variants: ring-rolling, open-die forging, closed-die forging, drop forging

Related processes in this database include: metal extrusion

Variants

Impression Die Forging—also called closed-die forging, presses metal between two dies that contain a precut profile of the desired part.

Cold Forging—includes bending, cold drawing, cold heading, coining, extrusions and more, to yield a diverse range of part shapes. The temperature of metals being cold forged may range from room temperature to several hundred degrees.

Open-die Forging is performed between flat dies with no precut profiles in the dies. Movement of the work-piece is the key to this method. Larger parts over 20 tones and 10 meters in length can be hammered or pressed into shape in this way.

Seamless Rolled Ring Forging is typically performed by punching a hole in a thick, round piece of metal (creating a donut shape), and then rolling and squeezing (or in some cases, pounding) the donut into a thin ring. Ring diameters can be anywhere from a few inches to 30 feet.

Process details

Closed-die forging

A heated blank is placed between two halves of a die (Figure 4-15).

A single compressive stroke squeezes the blank into the die to form the part. In *hammer* or *drop forging* this happens by dropping the top of the mould from a height. An alternative is to squeeze the moulds together using hydraulic pressure (Figure 4-16).

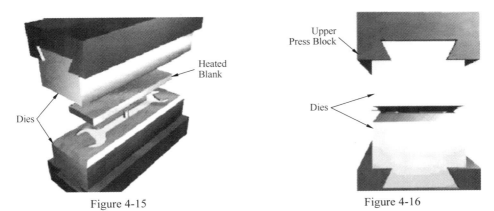

Figure 4-15   Figure 4-16

Once the die halves have separated, the part can be ejected immediately using an ejector pin

(Figure 4-17).

The waste material, flash, is removed later (Figure 4-18).

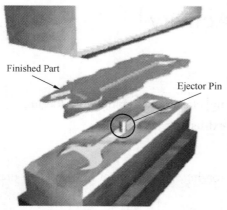

Figure 4-17

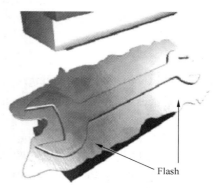

Figure 4-18

Materials and shapes

Any metal can be forged, provided the blank is hot enough (60% of the melting temperature). Typical possible sizes for closed dies range from 10g to 10kg, depending on complexity.

The part is left with good surface and mechanical properties, although cold-forging can perform even better. Complex parts can be formed using a series of forging dies with increasing levels of detail.

A draft (taper) angle has to be incorporated to allow easy removal of the part. Any waste material squeezed between the die halves, called flash, is readily recycled.

Unit 5
Introduction to CNC Machine and CAM Design

| Part I Technical and Practical Reading |

Passage A Basics of Computer Numerical Control

What is CNC?

CNC stands for Computer Numerical Control and has been around since the early 1970's. Prior to this, it was called NC, for Numerical Control. While people in most walks of life have never heard of this term, CNC has touched almost every form of manufacturing process in one way or another.

How CNC works

CNC machines are programmed by a CNC programmer, who uses a machining print to determine X, Y, and Z coordinates for each cutting tool inside the CNC to move to. This causes, the part that is loaded to be cut, drilled, tapped, bored, counter bored, chamfered, etc. Everything that an operator would be required to do with conventional machine tools is programmable with CNC machines. Once the machine is setup and running, a CNC machine is quite simple to keep running. In fact CNC operators tend to get quite bored during lengthy production runs because there is so little to do. With some CNC machines, even the work-piece loading process has been automated

(Figure 5-1).

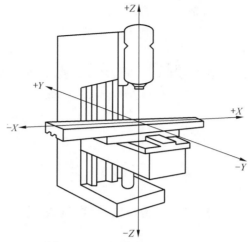

Figure 5-1 Mill Axes of Motion

Motion Control

All CNC machine types share this commonality: They all have two or more programmable directions of motion called axes. An axis of motion can be linear (along a straight line) or rotary (along a circular path). One of the first specifications that implies a CNC machine's complexity is how many axes it has. Generally speaking, the more axes, the more complex the machine.

In the beginning stepper motors, which rotate in increments, or steps, were the standard in motion control technology. Positional accuracy defines the precision with which a system can control the actual placement of the X, Y, and Z axes. Three-axis systems are the norm today, although some machines control five to seven axes of motion (Figure 5-2).

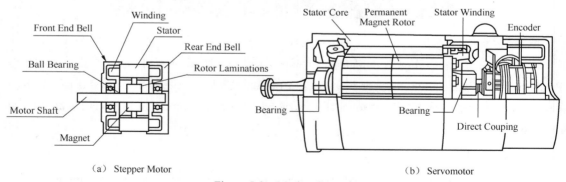

(a) Stepper Motor   (b) Servomotor

Figure 5-2 Motion Control

Programmable Accessories

A CNC machine wouldn't be very helpful if all it could only move the work-piece in two or more axes. Almost all CNC machines are programmable in several other ways. The specific CNC machine type has a lot to do with its appropriate programmable accessories. Here are some examples for one machine type.

Automatic Tool Changer

Most machining centers can hold many tools in a tool magazine. When required, the required tool can be automatically placed in the spindle for machining (Figure 5-3).

Spindle Speed and Activation

The spindle speed (in revolutions per minute) can be easily specified and the spindle can be turned on in a forward or reverse direction. It can also, of course, be turned off.

Coolant

Many machining operations require coolant for lubrication and cooling purposes. Coolant can be turned on and off within the machine cycle (Figure 5-4).

Figure 5-3 Automatic tool changer

Figure 5-4 Coolant

The CNC Program

A CNC program is nothing more than another kind of instruction set. It's written in sentence-like format and the control will execute it in sequential order, step by step.

A special series of CNC words are used to communicate what the machine is intended to do. CNC words begin with letter addresses (like F for feed-rate, S for spindle speed, and X, Y & Z for axis motion). When placed together in a logical method, a group of CNC words make up a command that resemble a sentence (Figure 5-5).

For any given CNC machine type, there will only be about 40-50 words used on a regular basis. So if you compare learning to write CNC programs to learning a foreign language having only 50 words, it shouldn't seem overly difficult to learn CNC programming.

The CNC control

The CNC control will interpret a CNC program and activate the series of commands in sequential order. As it reads the program, the CNC control will activate the appropriate machine functions, cause axis motion, and in general, follow the instructions given in the program. In. general, the CNC control allows all functions of the machine to be manipulated (Figure 5-6).

Figure 5-5　The CNC Programmming

Figure 5-6　The CNC Control

Notes

1. CNC machines are programmed by a CNC programmer, who uses a machining print to determine X, Y, and Z coordinates for each cutting tool inside the CNC to move to. 此句中 who 引导非限制性定语从句，先行词为 programmer。而 who 引导的定语从句中动词：uses 是谓语，不定式 to determine 做目的状语。整句可译为：计算机数控机床由数控编程员来编程，它用工艺文件来确定数控机床中每种切削工具移动的 X，Y 和 Z 坐标。

2. So if you compare learning to write CNC programs to learning a foreign language having only 50 words, it shouldn't seem overly difficult to learn CNC programming. 本句中 if 引导的条件状语从句中有两个 to。第一个 to 与 learn 构成 learn to do sth. "学习做某事"；第二个 to 与 compare 构成固定搭配，译为 "将……比作……"。本句译为：所以如果您将学习编写写计算机数控机床程序比作是学习一门只有50个单词的外语的话，那么计算机数控机床编程似乎并不十分难学。

3. As it reads the program, the CNC control will activate the appropriate machine functions, cause axis motion, and in general, follow the instructions given in the program. 这句中 as 引导时间状语从句。后面的主句中主语为 CNC control，谓语为三个动词：activate, cause, follow。本句译为：当读程序时，计算机数控机床的控制器会激活相应的机床功能、引起相应轴的运动，一般来说，在程序中还会跟随一系列的指令。

New Words

program ['prəugræm]　n. 程序
　　　　　　　　　　　v. 编程
programmer ['prəugræmə]　n. 编程员
coordinate [kəu'ɔ:dinit]　n. 坐标
chamfer ['tʃæmfə]　v. 倒角，倒槽
programmable ['prəugræməbl]　a. 可编程的
automate ['ɔ:təmeit]　v. 使自动化
commonality [,kɔmə'næliti]　n. 共性，通用性

Unit 5 Introduction to CNC Machine and CAM Design

linear ['liniə]   a. 线性的
complexity [kəm'pleksiti]   n. 复杂（性），复杂的事物
rotary ['rəutəri]   a. 旋转的
increment ['inkrimənt]   n. 增量
bell [bel]   n. 罩
winding ['waindiŋ]   n. 绕组
stator ['steitə]   n. 定子
rotor ['rəutə]   n. 转子
lamination [ˌlæmi'neiʃən]   n. 叠片，铁芯片
servomotor ['sə:vəuˌməutə]   n. 伺服电机
core [kɔ:]   n. 铁芯
encoder [in'kəudə]   n. 编码器
accuracy ['ækjurəsi]   n. 精度
placement ['pleismənt]   n. 位置
activation [ˌækti'veiʃən]   n. 启动，激活
revolution [ˌrevə'lu:ʃən]   n. 转数
specify ['spesifai]   v. 限定
reverse [ri'və:s]   a. 反的
coolant ['ku:lənt]   n. 冷却剂
lubrication [ˌlu:bri'keiʃən]   n. 润滑
purpose ['pə:pəs]   n. 作用
instruction [in'strʌkʃən]   n. 指令
overly ['əuvəli]   adv. 过度地，极度地
set [set]   n. 组
format ['fɔ:mæt]   n. 格式
execute ['eksikju:t]   v. 执行
sequential [si'kwinʃəl]   a. 顺序的
word [wə:d]   n. 指令字
feed-rate [fi:dreit]   n. 进给速度
activate ['æktiveit]   v. 激活，刺激
feedrate [fi:dəreit]   n. 进给速度
logical ['lɔdʒikəl]   a. 逻辑的
interpret [in'tə:prit]   v. 解译
manipulate [mə'nipjuleit]   v. 操纵

Phrases and Expressions

Computer Numerical Control (CNC)   计算机数字控制
manufacturing process   生产过程

counter bore 沉孔，锪沉孔
conventional machine 常规机床
stepper motor 步进电机
positional accuracy 位置精度
machine tool 机床
ball bearing 滚珠轴承
front end bell 前端罩
rear end bell 后端罩
rotor laminations 转子铁芯
direct coupling 直接联轴器
permanent magnet 永磁
revolutions per minute (RPM) 每分钟转数
automatic tool changer (ATC) 自动换刀装置
machining center 加工中心
tool magazine 刀库
spindle speed 主轴转速
sequential order 顺序次序
machine cycle 加工周期
axis of motion 运动轴

EXERCISE 1

The following is set of symbols denoting Working Safety. Choose the best symbol according to the information given below.

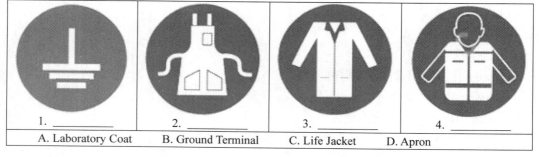

1. _____ 2. _____ 3. _____ 4. _____

A. Laboratory Coat   B. Ground Terminal   C. Life Jacket   D. Apron

EXERCISE 2

Translate the following phrases into Chinese or English.

1. Computer Numerical Control   _____
2. _____   常规机床
3. stepper motor   _____
4. _____   位置精度
5. machining center   _____
6. _____   主轴转速

7. direct coupling　　　　　　　_____
8. _____　　　　　　　自动换刀装置

Passage B Designing Parts with CAM Alibre

Alibre Design makes it easy to machine your parts with Alibre CAM. Based on MecSoft technology, Alibre CAM is included in Alibre's flagship product, Alibre Design Expert 10.0. Alibre CAM has a host of preset and customizable tools and output options. Supporting 2.5-axis, 3-axis, and drilling operations, Alibre CAM is a great tool for taking designs from concept to finished product.

Create a Part

Step 1: The first step in machining with Alibre CAM is to create a part. For this example, we'll use a part included in the Alibre CAM installation. Go to C:\Program Files\Alibre CAM 1.0\Tutorials and look for the part called 3Axis Example1.AD_PRT. Alternatively, you can create your own part to use (Figure 5-7).

Step 2: After you open your part in Alibre Design, proceed to the Alibre CAM Browser in

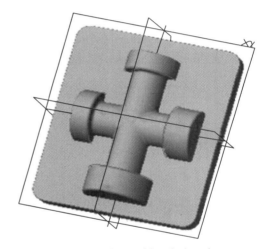

Figure 5-7　Start with a designed part

the menu bar at the top of the screen. Select Alibre CAM / Browser to activate the Alibre CAM interface (Figure 5-8). Once in Alibre CAM, click the Setup tab in the Alibre CAM Explorer. Click the Steup Machine button to change your initial tool position (Figure 5-9).

Figure 5-8　Select Alibre CAM

Figure 5-9　Select the Setup tab

Step 3: The Set Post-Processor Options menu lets you set batch options and select a program with which to view your machine code, if desired (Figure 5-10).

Step 4: The next step is to create your stock or load previously created stock. The Create / Load Stock and the Locate Part Within Stock buttons let you set the size of your initial stock material and locate your part within that stock (Figure 5-11).

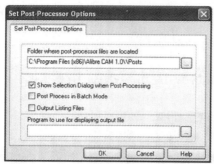

Figure 5-10   Choose the Set Post-Processor Options

Figure 5-11   Select the Create/Load Stock

Setup Tools

Now, we'll set up the tools. Click on the Tools tab to see some new options (Figure 5-12).

Step 5: Next, click the Create / Select Tool option. At the top of the Tool dialog box, you can choose from four standard tools (Figure 5-13). Click one of those tools and then set the properties and dimensions for your tool. When you have finished setting up the tool, click Save as New Tool and then OK to add your new tool to your list. As a side note, you also can load tool libraries or create your own to minimize rework.

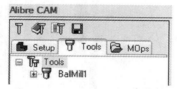

Figure 5-12   Choose the Tools tab

Figure 5-13   The Create/Select Tool button

You can adjust your tools in the Create/Select Tools dialog box (Figure 5-14).

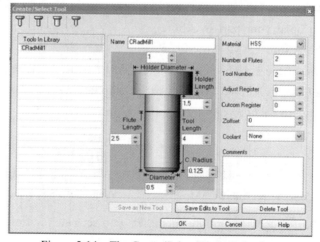

Figure 5-14   The Create/Select Tools dialog box

Add Machining Operations and Simulate the Tool-path

After we have our part, stock, and tool options set, it's time to set the type of machining that

Unit 5 Introduction to CNC Machine and CAM Design

we want to perform. In this example, we'll use parallel finishing as our machining type.

Step 6: With your tools in place, click on the MOps (short for Machining Operations) tab. Look for the Milling Methods button (Figure 5-15). When you click this button, a drop-down menu appears with the options Horizontal Roughing, Parallel Finishing, and Profiling.

Figure 5-15　Find the Milling Methods button

Step 7: Horizontal Roughing usually is applied first, and it generates the general shape of your part. Parallel Finishing usually is applied after Horizontal Roughing to add more definition to your shape by shaving off excess material left over from roughing. Click on the Parallel Finishing item, and the Parallel Finishing dialog box pops up (Figure 5-16). Set the options as appropriate for your application and then click Generate.

Next, you're ready to view the tool-path simulation. Click the Play button to watch the tool-path being created (Figure 5-17).

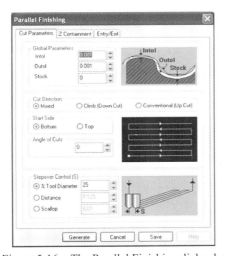

Figure 5-16　The Parallel Finishing dialog box

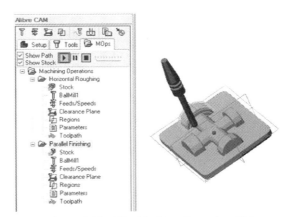

Figure 5-17　The Machine Operations Tab

Post-processing

After your tool-path is created, you must output it in the appropriate format depending on the machine you'll use. Alibre CAM comes with a host of postprocessors already defined for you to select from.

Step 8: Click the Post Processor button (Figure 5-18) and select your postprocessor. Set your output directory, press OK, and the file is generated.

You can use the output to machine your part.

Congratulations!You've just created your first machine code (Figure5-19) usingAlibre CAM.By exploring optionssuch as Feed/Speed settings, Clearance Control, Z-containment, and Approach and Engage values, you can fine tune your setup. If you need more tutorials or general help, you can always look in the Alibre CAM Help for more information and tips.

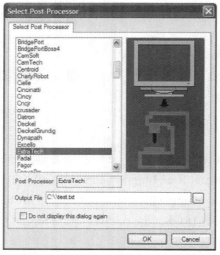
Figure 5-18 Select a postprocessor

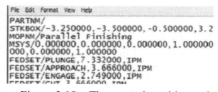
Figure 5-19 The created machine code

New Words

preset ['priːˈset]   v. / n. 预设
customizable [kʌstəmaizəbl]   a. 可定制化的
option [ˈɔpʃən]   n. 选项
installation [ˌinstəˈleiʃən]   n. 安装
alternatively [ɔːlˈtɜːnətivli]   ad. 作为选择
proceed [prəˈsiːd]   v. 继续
browser [brauzə(r)]   n. 浏览器
interface [ˈintə(ː)ˌfeis]   n. 界面
tab [tæb]   n. 标签
explorer [iksˈplɔːrə]   n. 资源管理器
initial [iˈniʃəl]   a. 最初的
post-processor [pəustˈprəusesə]   n. 后处理程序
batch [bætʃ]   n. 一组
view [vjuː]   v. 查看
save [seiv]   v. 保存
minimize [ˈminimaiz]   v. 减到最少
roughing [ˈrʌfiŋ]   n. 粗加工
profiling [ˈprəufailiŋ]   n. 设置文件
definition [ˌdefiˈniʃən]   n. 精确度
excess [ikˈses]   a. 多余的
simulation [ˌsimjuˈleiʃən]   n. 模拟
directory [diˈrektəri]   n. 目录
tune [tjuːn]   v. 调整

Unit 5 Introduction to CNC Machine and CAM Design

explore [iks'plɔː]　v. 探索
clearance ['kliərəns]　n. 间隙
containment [kən'teinmənt]　n. 控制
approach [ə'prəutʃ]　n. 接近
engage [in'geidʒ]　n. 啮合
tutorial [tjuː'tɔːriəl]　n. 教程

Phrases and Expressions

Computer-assisted Manufacture （CAM）　计算机辅助制造
flagship product　龙头产品
a host of　大量的
finished product　成品
menu bar　菜单栏
machine code　机器代码
dialog box　对话框
tool library　刀库
tool path　刀具路径
machining type　加工类型
drop-down menu　下拉菜单
shave off　除去
pop up　弹出
fine tune　微调

EXERCISE 3

Choose the control buttons of CNC machines according to the information given.

1.　A. Switch between machining or programming modes
2.　B. Manual operation
3.　C. Positioning with MDI
4.　D. Switching the soft-key rows
5.　E. Automatic (AUTO)
6.　F. Setting the screen layout
7.　G. File names/Comments

8.   H. Electronic hand-wheel

9.   I. Enter and call tool length and radius

10.   J. Single block

EXERCISE 4

Abbreviations are very useful in practical work. Read them and then translate them into corresponding Chinese terms.

1. IT	Interfacial Tension	_____
2. L.B.	Local Battery	_____
3. LD	Leak Detector	_____
4. KE	Kinetic Energy	_____
5. LOA	Length Over All	_____
6. mail. i	Malleable Iron	_____
7. man. op	Manual Operation	_____
8. MD	Maintenance Division	_____

Part II Glance at Conventional Machine Tool Structures

The following is the structure of CNC mill machine

DM45 3-Axis Tool-room Bed Mill

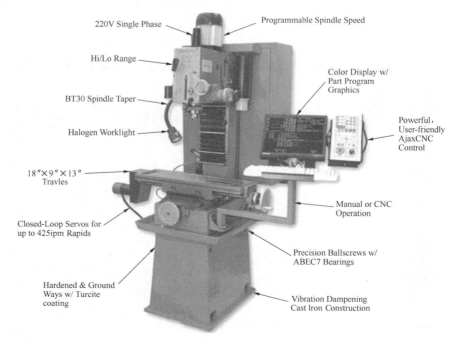

Unit 5 Introduction to CNC Machine and CAM Design

Explanations of the CNC Mill Machine Terms

No.	Name	Explanation
1	220v Single Phrase	220V 单相电源
2	Hi/Lo Range	高/低电压控制旋钮
3	BT30 Spindle Taper	BT30 主轴锥度
4	Halogen Work-light	Halogen 照明灯
5	18"×9"×13"Travels	行程范围 18"×9"×13"
6	Closed-Loop Servos for up to 425ipm Rapids	闭环伺服快速进给 425 脉冲/分
7	Hardened &Ground Ways w/Turcite Coating	已硬化且磨光的滑道/耐磨镀层
8	Vibration Dampening Cast Iron Construction	减震铸铁结构
9	Precision Ball-screws w/ABEC7 Bearings	精密滚珠丝杠/ABEC7 轴承
10	Manual or CNC Operation	手动或 CNC 操作
11	Powerful User-friendly AjaxCNC Control	功能强大的用户友好界面 AjaxCNC 控制
12	Color Display w/Part Program Graphics	彩色显示器/显示零件程序图形
13	Programmable Spindle Speed	主轴速度可编程

EXERCISE 5

The following is a CNC machine tool. You are required to choose the suitable words or phrases given below.

| Surface- Grinding Machine | Centerless Grinding Machine | Internal Grinding Machine |
| Jig Grinding Machine | Slideway Grinding Machine | Too Grinding Machine |

Grinding Machine（磨床）

1. _____（内圆磨床）

2. _____（平面磨床）

3. _____（导轨磨床）

续表

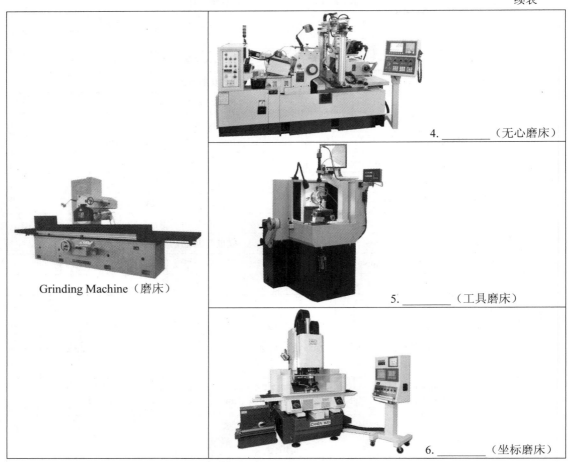

Grinding Machine（磨床）

4. _____（无心磨床）

5. _____（工具磨床）

6. _____（坐标磨床）

| Part III Simulated Writing |

Section A Match Your Skill

The following is panel of a CNC machine, and you can understand the functions of its buttons.

Unit 5 Introduction to CNC Machine and CAM Design

Explanations of the CNC Panel

No.	Name	Explanation
1	Alphabetic keyboard for entering texts and file names, as well as for programming in ISO format	字母键盘用于输入文本及文件名称，也可按 ISO 格式进行编程
2	• File management • Pocket calculator (not TNC 410) • MOD functions • HELP functions	• 文件管理 • 便携式计算器（并不是 TNC 410 型） • 模式功能 • 辅助功能
3	Programming modes	编程模式
4	Machine operating modes	机床操作模式
5	Initiation of programming dialog	可进行对话编程
6	Arrow keys and GOTO jump command	箭头键和跳转指令
7	Numerical input and axis selection	数字输入和轴选择

EXERCISE 6

Match the words or phrases on the left with their meanings on the right.

Automatic Tool Magazine

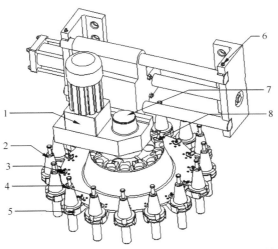

1. Transmission
2. Tool handles
3. Positioning key for tool handle
4. Blind rivet
5. Tool apron
6. Fastening screw for guide rail
7. Rotor gland
8. Indexing Geneva wheel

A．刀柄定位键
B．刀座
C．导轨紧固螺钉
D．变速器
E．分度槽轮
F．刀柄
G．拉钉
H．转轴压盖

Section B Have a Try

This section will help you to understand several forms of machining.

External Operations

Thread cutting—A single-point threading tool, typically with a 60 degree pointed nose, moves axially, along the side of the work-piece, cutting threads into the outer surface. The threads can be cut to a specified length and pitch and may require multiple passes to be formed.

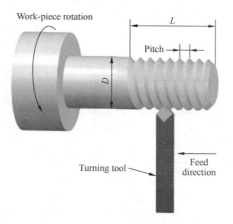

EXERCISE 7

This section is to test your ability to identify different operations.

| Ball end mill | Tool nose radius | Chamfer mill |

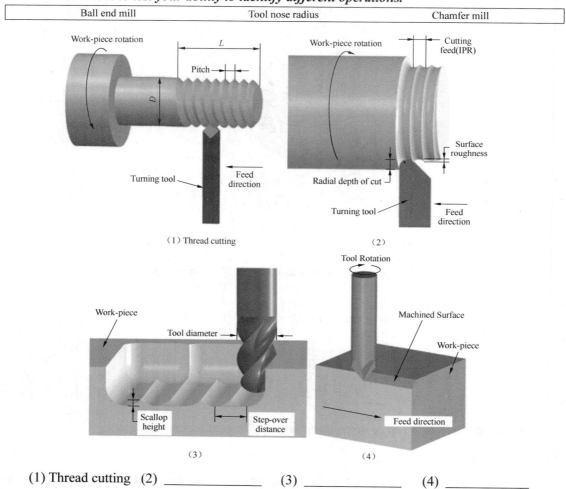

(1) Thread cutting (2) _____ (3) _____ (4) _____

Part IV Broaden Your Horizon—Practical Activity

Kow to Produce New Firebox Door Patterns

These are two of the old foundry patterns for our firebox doors. They are epoxy castings mounted in a plywood plate (Figure 5-20). The foundry switched to a new high pressure sand molding process, and the patterns broke (Figure 5-21). We needed higher strength aluminum patterns.

Figure 5-20 Old foundry patterns

Figure 5-21 New door to be designed

Because the old patterns were cast from a hand-made master, there were also problems such as dimensional inconsistencies, etc. The first step in building new patterns was to have a new vertical CNC milling machine (Figure 5-22).

Figure 5-22 CNC Milling Machine

Next, the doors were modeled in a Rhino, a new 3-D modeller. Here's a screenshot of the Rhino model (Figure 5-23).

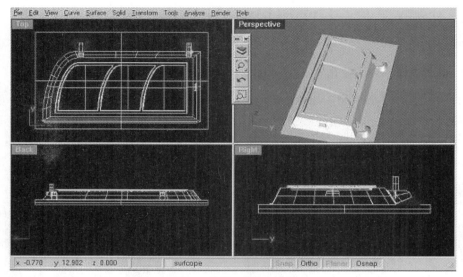

Figure 5-23   3D model

Here's a close-up of the parting line around the bottom hinge lug (Figure 5-24).
Here's a new feature in Rhino 1.1 Beta (Aug 20): Draft Angle Analysis (Figure 5-25).

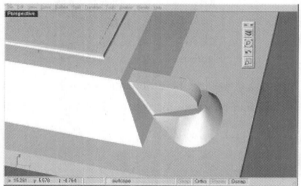

Figure 5-24   Figure 5-25

Next, the 3-D Rhino model was imported into Visual Mill, a new CAM package that generates the cutter paths for the milling machine directly from Rhino NURBS models. Here is a screen shot of Visual Mill simulating the cutting of an aluminum block (Figure 5-26).

Here's another screen shot showing the tool-paths for a 1/8" ball shaped cutter as it goes into the corners around the hinges to clean up the leftovers from the previous larger cutter (Figure 5-27).

Tool-paths were loaded onto a floppy disk and transferred to the mill's controller. First, they were tested on a block of wax (Figure 5-28).

Next, they were cut from a 2 1/4" thick block of aluminum. It takes about 12 hours of machine time to cut each side of a two-sided pattern.

Here's a photo of the finished patterns and their mounting plates (Figure 5-29).

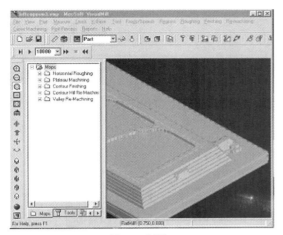

Figure 5-26

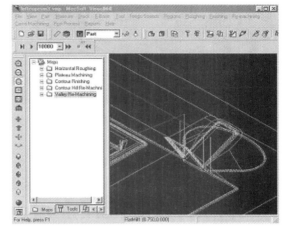

Figure 5-27

Figure 5-28

Figure 5-29

Figure 5-30

Unit 6
Engineering Drawings

| Part I   Technical and Practical Reading |

Passage A   Technical Drawing (I)

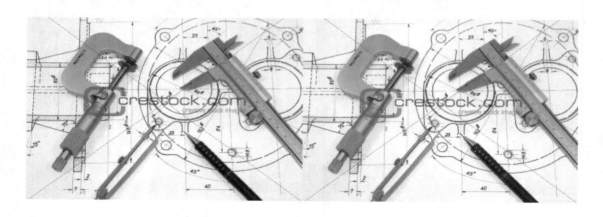

Technical drawing, also known as drafting, is the academic discipline of creating standardized technical drawings by architects, interior designers, drafters, design engineers, and related professionals. Standards and conventions for layout, line thickness, text size, symbols, view projections, descriptive geometry, dimensioning, and notation are used to create drawings that are ideally interpreted in only one way.

A person who does drafting is known as a drafter. In some areas this person may be referred to as a drafting technician, draftsperson, or draughts person. This person creates technical drawings which are a form of specialized graphic communication (Figure 6-1). A technical drawing differs from a common drawing by how it is interpreted. A common drawing can hold many purposes and meanings, while a technical drawing is intended to concisely and clearly communicate all needed specifications to transform an idea into physical form (Figure 6-2).

Unit 6 Engineering Drawings

Figure 6-1   Drafter at work

Figure 6-2   Copying technical drawings

Types of Technical Drawings

Engineering drawing

Engineering drawing is a type of technical drawing, created within the engineering discipline, and used to fully and clearly define requirements for engineered items.

Engineering drawings are usually created in accordance with standardized conventions for layout, nomenclature, interpretation, appearance (such as typefaces and line styles), size, etc (Figure 6-3).

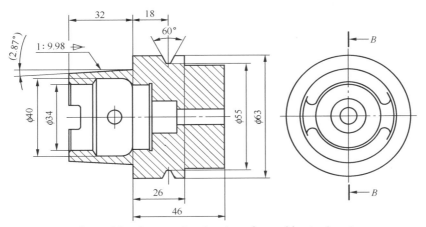
Figure 6-3   Construction drawing of a machine tool part

Its purpose is to accurately and unambiguously capture all the geometric features of a product or a component. The end goal of an engineering drawing is to convey all the required information that will allow a manufacturer to produce that component.

Cutaway drawings

A cutaway drawing is a technical illustration, in which surface elements of a three-dimensional model are selectively removed, to make internal features visible, but without sacrificing the outer context entirely (Figure 6-4).

The purpose of a cutaway drawing is to allow the viewer to have a look into an otherwise solid opaque object. Instead of letting the inner object shine through the surrounding surface, parts of outside object are simply removed. This produces a visual appearance as if someone had cutout a piece of the object or sliced it into parts. Cutaway illustrations avoid ambiguities with respect to spatial ordering, provide a sharp contrast between foreground and background objects, and facilitate a good understanding of spatial ordering.

Figure 6-4　Cutaway drawing of a Nash 600

Exploded view drawing

An exploded view drawing is a technical drawing of an object that shows the relationship or order of assembly of the various parts. It shows the components of an object slightly separated by distance, or suspended in surrounding space in the case of a three-dimensional exploded diagram. An object is represented as if there had been a small controlled explosion emanating from the middle of the object, causing the object's parts to be separated an equal distance away from their original locations (Figure 6-5).

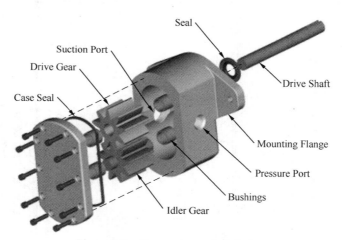

Figure 6-5　Gear pump exploded view

An exploded view drawing can show the intended assembly of mechanical or other parts. In mechanical systems usually the component closest to the center is assembled first, or is the main part in which the other parts get assembled. This drawing can also help to represent disassembly of parts, where the parts on the outside normally get removed first.

Patent drawing

A patent drawing is a technical drawing of a patent invention, which shows the nature of the invention. The drawing must show every feature of the invention specified in the claims, and is required by the patent office rules to be in a particular form (Figure 6-6).

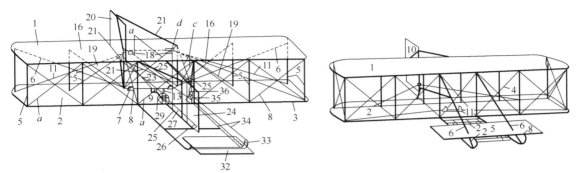

Figure 6-6　Wright brother's patent drawing

The applicant for a patent will be required by law to furnish a drawing of the invention whenever the nature of the case requires a drawing to understand the invention. This drawing must be filed with the application. This includes practically all inventions except compositions of matter or processes, but a drawing may also be useful in the case of many processes.

Notes

1. Technical drawing, also known as drafting, is the academic discipline of creating standardized technical drawings by architects, interior designers, drafters, design engineers, and related professionals. "also known as drafting" 为插入语。而 "architects, interior designers… and related professionals" 是并列结构。全句可译为：技术制图，又称制图，是由建筑师、室内设计师、制图员、设计工程师以及相关专业人士用来创建标准技术图纸的学科。

2. The applicant for a patent will be required by law to furnish a drawing of the invention whenever the nature of the case requires a drawing to understand the invention. 由 whenever 引导出时间状语从句。本句可译为：当需要通过图纸来理解发明的性质时，法律要求专利申请人提供发明图纸。

New Words

drafting ['drɑːftiŋ]　n. 制图
academic [ˌækə'demik]　a. 学术的
discipline ['disiplin]　n. 科目
drafter ['drɑːftə]　n. 制图员
convention [kən'venʃən]　n. 常规
layout ['leiˌaut]　n. 布局

projection [prə'dʒekʃən]   n. 投影
dimensioning [di'menʃən]   n. 标注尺寸
notation [nəu'teiʃən]   n. 记号
technician [tek'niʃ(ə)n]   n. 技术员
draftsperson [drɑ:fts'pə:sn]   n. 绘图员
graphic ['græfik]   a. 图形的
physical ['fizikəl]   a. 实物的
nomenclature [nəu'menklətʃə]   n. 术语
typeface ['taipfeis]   n. 字型
illustration [ˌiləs'treiʃən]   n. 图解
three-dimensional [θri:'dimenʃənəl]   a. 三维的
context ['kɔntekst]   n. 背景
solid ['sɔlid]   a. 实体的
opaque [əu'peik]   a. 不透明的
cutout ['kʌtaut]   v. 切断、切除
slice [slais]   v. 切
spatial ['speiʃəl]   a. 空间的
foreground ['fɔ:graund]   n. 前景
background ['bækgraund]   n. 背景
suspend [səs'pend]   v. 悬置
represent [ˌri:pri'zent]   v. 表示
emanate ['eməneit]   v. 发源，产出
location [ləu'keiʃən]   n. 位置
disassembly [ˌdisə'sembli]   n. 拆卸
remove [ri'mu:v]   v. 拆卸
process [prə'ses]   n. 工艺

Phrases and Expressions

engineering drawing   工程图
technical drawing   技术图
academic discipline   学科
line thickness   线的粗细
text size   文本长度
view projection   视图、投影图
descriptive geometry   画法几何
draughts person   制图员
physical form   实物形式
engineered item   工程部件

be in accordance with   依据
line style   线型
geometric feature   几何特征
cutaway drawing   剖视图
technical illustration   技术图解
three-dimensional model   三维模型
visual appearance   视觉外观
spatial ordering   空间排序
exploded view drawing   分解图
patent drawing   专利图

EXERCISE 1

The following is set of symbols denoting Working Safety. Choose the best symbol according to the information given below.

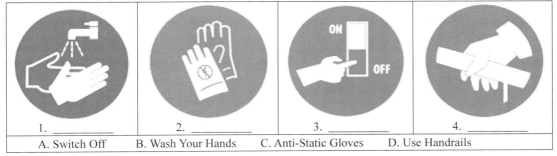

1. _____   2. _____   3. _____   4. _____
A. Switch Off   B. Wash Your Hands   C. Anti-Static Gloves   D. Use Handrails

EXERCISE 2

Translate the following phrases into Chinese or English.

1. view projection _____   2. _____ 工程图
3. cutaway drawing _____   4. _____ 画法几何
5. patent drawing _____   6. _____ 三维模型
7. exploded view drawing _____ 8. _____ 视觉外观

Passage B   Technical Drawing (II)

Technical illustrations

Technical illustration is the use of illustration to visually communicate information of a technical nature. Technical illustrations can be component technical drawings or diagrams. The aim of technical illustration is to generate expressive images that effectively convey certain information via the visual channel to the human observer (Figure 6-7).

The main purpose of technical illustration is to describe or explain these items to a more or less non-technical audience. The visual image should be accurate in terms of dimensions and proportions, and should provide an overall impression of what an object is or does, to enhance the viewer's interest and understanding.

Technical sketches

A sketch is a rapidly executed freehand drawing that is not intended as a finished work. In general, a sketch is a quick way to record an idea for later use. Architect's sketches primarily serve as a way to try out different ideas and establish a composition before undertaking a more finished work, especially when the finished work is expensive and time consuming (Figure 6-8).

Figure 6-7 Technical illustrations   Figure 6-8 Sketches for a government building

Architectural sketches, for example, are a kind of diagrams. These sketches, like metaphors, are used by architects as a mean of communication in aiding design collaboration.

Basic drafting paper sizes

As can be seen in the series, the height of the previous drawing size becomes double the height of the next size in the sequence (Table 6-1).

Table 6-1　　　　　　　　　　Basic drafting paper sizes

Drawing type "name"	Dimensions (width × height)	Drawing type "name"	Dimensions (width × height)
A-size	8.5 by 11.0 inches 22 cm by 28 cm	B-size	11.0 by 17.0 inches 28 cm by 43 cm
C-size	17.0 by 22.0 inches 43 cm by 56 cm	D-size	22.0 by 34.0 inches 56 cm by 86 cm
E-size	34.0 by 44.0 inches 86 cm by 112 cm	F-size	44.0 by 68.0 inches 112 cm by 173 cm
G-size	68.0 by 88.0 inches 173 cm by 224 cm	H-size	88.0 by 136 inches 224 cm by 345 cm

Manual drafting

The basic drafting procedure is to place a piece of paper on a smooth surface with right-angle corners and straight sides—typically a drafting table (Figure 6-9). A sliding straightedge known as a T-square is then placed on one of the sides, allowing it to be slid across the side of the table, and over the surface of the paper.

"Parallel lines" can be drawn simply by moving the T-square and running a pencil or technical pen along the T-square's edge, but more typically the T-square is used as a tool to hold other devices such as set squares or triangles. Modern drafting tables (which have by now largely been replaced by CAD workstations) come equipped with a parallel rule that is supported on both sides of the table to slide over a large piece of paper.

Figure 6-9   A drafting table

In addition, the drafter uses several tools to draw curves and circles. Primary among these are the compasses, used for drawing simple arcs and circles, and the French curve, typically a piece of plastic with complex curves on it. A spline is a rubber-coated articulated metal that can be manually bent to most curves (Figure 6-10).

Figure 6-10   Technical drawing instrument

Computer Aided Design

Today, the mechanics of the drafting task have largely been automated and accelerated through the use of Computer Aided Design (CAD). systems Computer-aided design is the use of computer technology to aid in the design and particularly the drafting of a part or product, including entire buildings (Figure 6-11).

Drafting can be done in two dimensions ("2D") and three dimensions ("3D"). Drafting is the integral

Figure 6-11   Computer Aided Design

communication of technical or engineering drawings and is the industrial arts sub-discipline. In representing complex, three-dimensional objects in two-dimensional drawings, these objects have traditionally been represented by three projected views at right angles.

New Words

margin ['mɑ:dʒin]　n. 页边距
visually ['vizjuəli]　ad. 可视地，形象化地
proportion [prə'pɔ:ʃən]　n. 比例
sketch [sketʃ]　n. 草图、素描
freehand ['fri:hænd]　a. 凭手画的
architectural [,ɑ:ki'tektʃərəl]　a. 建筑的
series ['siəri:z]　n. 系列
sequence ['si:kwəns]　n. 次序
procedure [prə'si:dʒə]　n. 步骤
straightedge ['streitedʒ]　n. 直尺
T-square [ti:-skweə]　n. 丁字尺
triangle ['traiæŋgl]　n. 三角板
workstation ['wə:ksteiʃ(ə)n]　n. 工作站
curve [kə:v]　n. 曲线
arc [ɑ:k]　n. 弧
size [saiz]　n. （图纸）号
compass ['kʌmpəs]　n. 圆规
spline [splain]　n. 曲线规
rubber ['rʌbə]　n. 橡胶
articulated [ɑ:'tikjulitid]　a. 铰链的
manually ['mænjuəli]　ad. 人工地
mechanics [mi'kæniks]　n. 技巧
accelerate [æk'seləreit]　v. 加速
sub-discipline ['sʌb'disiplin]　n. 分支学科

Phrases and Expressions

in terms of　在……方面
freehand drawing　徒手图
design collaboration　协同设计
drafting paper　绘图纸
manual drafting　手工制图
right angle　直角

straight side　直边
drafting table　制图台
parallel line　平行线
set square　三角板
technical pen　专业碳素笔
Computer Aided Design (CAD)　计算机辅助设计
parallel rule　平行规
French curve　曲线板
two dimensions (2D)　二维
three dimensions (3D)　三维
industrial arts　工艺美术
projected view　投影视图

EXERCISE 3

Match the step with the picture given.

Quick Tool Change

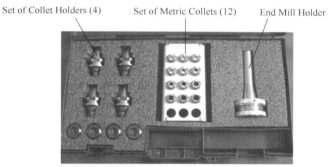

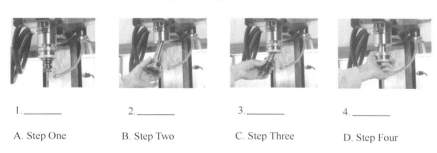

1._____　　2._____　　3._____　　4._____

A. Step One　　B. Step Two　　C. Step Three　　D. Step Four

EXERCISE 4

Abbreviations are very useful in practical work. Read them and then translate them into corresponding Chinese terms.

1. ME	Mechanical Engineer	_____
2. FWB	Flexible Wheel Base	_____

3. M.O.	Master Oscillator	_____
4. M.S.	Machinery Steel	_____
5. MS	Margin of Safety	_____
6. MU	Measurement Unit	_____
7. NA	Neutral Axis	_____
8. OS	Operating System	_____

Part II　Glance at Technical Drawing Instruments

The following are some drawing instruments.

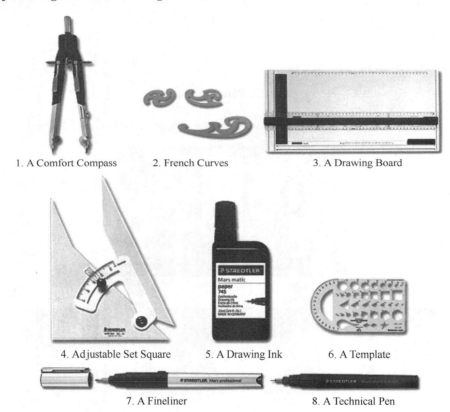

1. A Comfort Compass　　2. French Curves　　3. A Drawing Board

4. Adjustable Set Square　　5. A Drawing Ink　　6. A Template

7. A Fineliner　　8. A Technical Pen

Explanations of the Drawing Instrument Terms

No.	Name	Explanation
1	A Comfort Compass	圆规
2	French Curves	曲线板
3	A Drawing Board	绘图板
4	Adjustable Set Square	可调式三角板斜角规
5	A Drawing Ink	绘图墨水
6	A Template	样板
7	A Fineliner	细芯笔
8	A Technical Pen	专业碳素笔

Unit 6 Engineering Drawings

EXERCISE 5

The following are some drawing instruments. You are required to choose the suitable name for each instrument.

| An Automatic Pencil | Highlighter | An Ink Pen | A Rapid Eraser |
| A Graphite Pen | A Graphite Pencil | A Comfort Compass | |

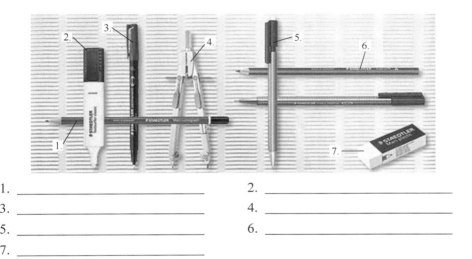

1. _____   2. _____
3. _____   4. _____
5. _____   6. _____
7. _____

| Part III   Simulated Writing |

Section A   Match Your Skill

The following is a lathe accessory, collet chuck adaptor, and you can learn the names of its parts.

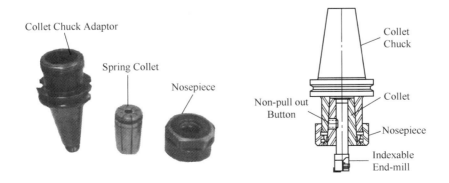

Explanations of Terms

No.	Name	Explanation
1	Collet Chuck Adaptor	筒夹接头
2	Spring Collet	弹簧夹头
3	Nosepiece	喷嘴
4	Collet Chuck	筒夹
5	Non-pull out Button	按钮
6	Collet	夹头
7	Indexable End-mill	可转位铣头

EXERCISE 6

Match the words or phrases on the left with their meanings on the right.

Universal Boring and Facing Head

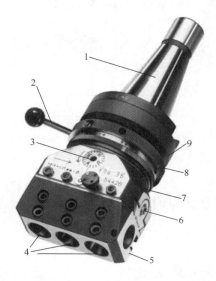

1. Exchangeable taper shank   A. 螺纹标度
2. Stopping bar   B. 滑块
3. Worm with scale   C. 刀身
4. Modular clamping   D. 可替换锥柄
5. Slide   E. 安全销
6. Adjusting screw with scale   F. 止动杆
7. Body   G. 模块夹具系统
8. Braking ring   H. 标度调整螺钉
9. Safety pin   I. 制动环

Section B Have a Try

This section will help you to understand several forms of machining

Internal Operations

Drilling—An operation in which a drill enters the work-piece axially and cuts a hole with a diameter equal to that of the tool. On a milling machine, an end milling operation is required to produce a hole with a tool smaller than the hole diameter. A drilling operation typically produces a blind hole, which extends to some depth inside the work-piece, measured to the point made by the tool or to the end of the full diameter portion. On a milling machine, a hole that extends completely through the work-piece (through hole) can also be drilled.

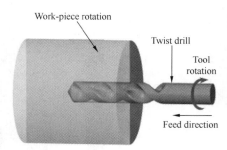

Unit 6 Engineering Drawings

EXERCISE 7

This section is to test your ability to identify different operations.

Through hole   Blind hole   Drilling (Milling machine)

(1) Drilling

(2)

(3)

(4)

(1) Drilling (Turning machine) (2) _____
(3) _____   (4) _____

| Part IV   Broaden Your Horizon—
Practical Activity |

Assembly of Collet

Loading your tools properly in the collets and tool holders is a critical step in setting up the machine and should be done carefully to reduce the risk of breaking a tool and/or getting injured.

1. Begin by choosing a collet matching the diameter of your tool such that the shank of the tool easily slides into the hole.

2. The collet and nut need to be assembled before mounting to the tool holder (Figure 6-12).

3. You will see that there is a small groove on

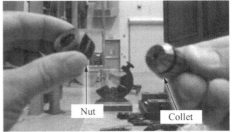

Figure 6-12   Collet and nut assembly

| 111 |

the thicker end of the collet. This groove must be snapped into the nut before mounting to the tool holder. To do this, push the collet into the back of the nut at a slight angle, then twist it and align it with the axis of the nut until you hear it "click" into place. As such, it should rotate freely inside the nut (Figure 6-13).

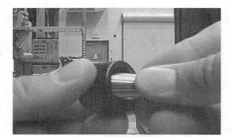 

Figure 6-13　Inserting the collet

4. At this point it is a good idea to check to see that your tool easily slides into the hole in the collet. If not, you might have the wrong collet, or you did not properly snap it into place (Figure 6-14).

5. Place the tool holder into the fixture. You need to align the slots in the tool holder with the tabs in the fixture. Also, the tool holder will only go into the fixture in one orientation. The slots are slightly different in width, as are the tabs on the fixture. You can easily figure out the orientation, by facing the slot with the small tapered hole inside it toward you (Figure 6-15).

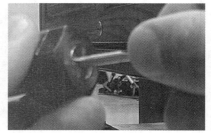

Figure 6-14　Checking for clearance

（a）Insert tool holder　　　　　　　　（b）Align tool holder

Figure 6-15

6. Now you can take the collet and nut and slightly screw it onto the tool holder. (You may need to do this sideways in order to keep the collet from falling back out of the nut, and then put the assembled tool holder back in the fixture.) (Figure 6-16)

7. Place the tool shank into the collet (If it seems too tight, you may have screwed the nut on too much). You'll need to hold the tool with the shank in the collet at a proper depth, while hand-tightening the nut (Figure 6-17).

(a) Place collet and nut　　　(b) Tighten slightly

　　　　Figure 6-16　　　　　　　　　　　　　　Figure 6-17　Insert tool

8. Finally, use a wrench and tighten the nut down. This needs to be tight, but you can damage the collet if you over-tighten.

Unit 7
Electronic Components and Circuit

| Part I Technical and Practical Reading |

Passage A Electronic Components and Symbols

An electronic component is a basic electronic element and is usually packaged in a discrete form with two or more connecting leads or metallic pads. Components are intended to be connected together, usually by soldering to a printed circuit board, to create an electronic circuit with a particular function (for example an amplifier, radio receiver, or oscillator).

There are a large number of symbols which represent an equally large range of electronic components. It is important that you can recognize the more common components and understand what they actually do.

Diodes

Diodes are basically a one-way valve for electrical current. They let it flow in one direction (from positive to negative) and not in the other direction. Most diodes are similar in appearance to a resistor and will have a painted line on one end showing the direction or flow (white side is negative). If the negative side is on the negative end of the circuit, current will flow. If the negative is on the positive side of the circuit, no current will flow (Figure 7-1).

Transistors

There are two types of standard transistors, NPN and PNP, with different circuit symbols. The transistor is possibly the most important invention of this decade. It performs two basic functions. (1) It acts as a switch turning current on and off. (2) It acts as an amplifier. This makes an output signal that is a magnified version of the input signal (Figure 7-2).

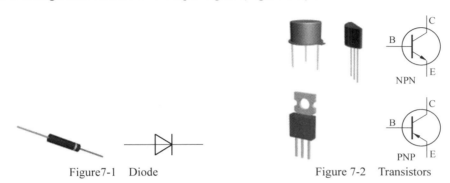

Figure 7-1　Diode　　　　　Figure 7-2　Transistors

Resistors

Resistors, like diodes and relays, are another of the electronic parts that should have a section in the installer's parts bin. They have become a necessity for the mobile electronics installer, whether it be for door locks, timing circuits, remote starts, LED's, or just to discharge a stiffening capacitor (Figure 7-3).

Resistors "resist" the flow of electrical current. The higher the value of resistance (measured in ohms) the lower the current will be.

Capacitors

A device used to store charge in an electrical circuit. A capacitor functions much like a battery, but charges and discharges much more efficiently (batteries, though, can store much more charge) (Figure 7-4).

Figure 7-3　Resistors　　　　　Figure 7-4　Capacitors

A basic capacitor is made up of two conductors separated by an insulator, or dielectric.

Thyristor

The thyristor is a solid-state semiconductor device with four layers of alternating N and P-type material. They act as bistable switches, conducting when their gate receives a current pulse, and continue to conduct for as long as they are forward biased (that is, as long as the voltage across the device has not reversed) (Figure 7-5).

Some sources define silicon controlled rectifiers and thyristors as synonymous.

Light-emitting Diode

A light-emitting diode (LED) is an electronic light source. LEDs are based on the semiconductor diode. When the diode is forward biased (switched on), electrons are able to recombine with holes and energy is released in the form of light. This effect is called electroluminescence and the color of the light is determined by the energy gap of the semiconductor (Figure 7-6).

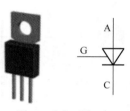

Figure 7-5　Thyristor

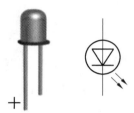

Figure 7-6　Light "-" emitting Diode

LEDs present many advantages over traditional light sources including lower energy consumption, longer lifetime, improved robustness, smaller size and faster switching.

Relay

A relay is an electrically operated switch. Current flowing through the coil of the relay creates a magnetic field which attracts a lever and changes the switch contacts. The coil current can be on or off so relays have two switch positions and they are double throw (changeover) switches (Figure 7-7).

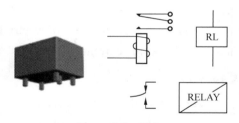

Figure 7-7　Relay

Notes

1. It is important that you can recognize the more common components and understand what they actually do. 本句为主语从句。句中 recognize 和 understand 为并列结构；而 understand 又引导出一个宾语从句。全句可译为：认识常见的部件并了解它们的功用是非常重要的。

2. The coil current can be on or off so relays have two switch positions and they are double throw (changeover) switches. 本句为并列句，由并列连词 so 引出。they 指代 relays。全句可译为：线圈中可允许或阻止电流流过，所以继电器可处于开关两个位置，作为双向开关 (转换开关)。

Unit 7 Electronic Components and Circuit

New Words

electronic [ilek'trɔnik]   a. 电子的
component [kəm'pəunənt]   n. 部件
element ['elimənt]   n. 组件
package ['pækidʒ]   v. 封装
discrete [dis'kri:t]   a. 离散的
metallic [mi'tælik]   a. 金属的
pad [pæd]   n. 垫片
solder ['sɔldə]   v. 焊接
amplifier ['æmpli,faiə]   n. 放大器
oscillator ['ɔsileitə]   n. 振荡器
symbol ['simbəl]   n. 符号
diode ['daiəud]   n. 二极管
positive ['pɔzətiv]   n. 正极
negative ['negətiv]   n. 负极
resistor [ri'zistə]   n. 电阻器
transistor [træn'zistə]   n. 晶体管
magnify ['mægnifai]   v. 放大
relay ['ri:lei]   n. 继电器
mobile ['məubail]   a. 移动的
electronics [ilek'trɔniks]   n. 电子学
remote [ri'məut]   a. 远程的
stiffen ['stifn]   v. 硬化
capacitor [kə'pæsitə]   n. 电容器
resistance [ri'zistəns]   n. 电阻(值)
ohm [əum]   n. 欧姆
charge [tʃɑ:dʒ]   n. 电荷 v. 充电
conductor [kən'dʌktə]   n. 导体
insulator ['insjuleitə]   n. 绝缘体
dielectric [,daii'lektrik]   n. 电介质
thyristor [θai'ristə]   n. 晶闸管
solid-state ['sɔlidsteit]   a. 固态的
semiconductor ['semikən'dʌktə]   n. 半导体
alternating [ɔ:l'tə:nitiŋ]   a. 交替的
bistable [bai'steibl]   a. 双稳态的
gate [geit]   n. 门极
bias ['baiəs]   v. 偏置

rectifier [ˈrektifaiə]   n. 整流器
emit [iˈmit]   v. 发射
electron [iˈlektrɔn]   n. 电子
recombine [ˌri:kəmˈbain]   v. 重新组合
electroluminescence [iˈlektrəuˌlju:miˈnesəns]   n. 电致发光
consumption [kənˈsʌmpʃən]   n. 消耗
robustness [rəˈbʌstnis]   n. 强度
coil [kɔil]   n. 线圈
contact [ˈkɔntækt]   n. 触点
changeover [ˈtʃeindʒˈəuvə]   n. 转换

Phrases and Expressions

printed circuit board (PCB)   印刷电路板
a large range of   大量的，大范围的
act as   作为
timing circuit   定时电路
light-emitting diode (LED)   发光二极管
be made up of   由……组成
define…as…   把……定义为……
silicon controlled rectifier   可控硅整流器
light source   光源
in the form of   以……形式
energy gap   能隙
energy consumption   能耗
magnetic field   磁场
double throw switch   双向开关

EXERCISE 1

The following is set of symbols denoting Working Safety. Choose the best symbol according to the information given below.

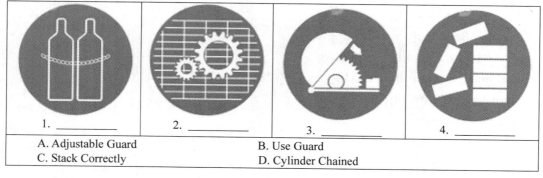

1. ____ 2. ____ 3. ____ 4. ____

A. Adjustable Guard   B. Use Guard
C. Stack Correctly   D. Cylinder Chained

Unit 7 Electronic Components and Circuit

EXERCISE 2

Translate the following phrases into Chinese or English.

1. electronic components _____
2. _____ 金属垫片
3. one-way valve _____
4. _____ 电流
5. current pulse _____
6. _____ 半导体装置
7. bistable switches _____
8. _____ 发光二极管

Passage B Basic Circuit Concepts

The figure shows the basic type of electrical circuit (Figure 7-8), in the form of a block diagram. It consists of a source of electrical energy, some sort of load to make use of that energy, and electrical conductors connecting the source and the load.

The electrical source has two terminals, designated positive (+) and negative (−). As long as there is an unbroken connection from source to load and back again as shown here, electrons will be pushed from the negative terminal of the source, through the load, and then back to the positive

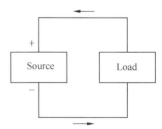

Figure 7-8 Electrical circuit

terminal of the source. The arrows show the direction of electron current flow through this circuit. Because the electrons are always moving in the same direction through the circuit, their motion is known as a *direct current* (DC).

The *source* can be any source of electrical energy. In practice, there are three general possibilities: it can be a battery, an electrical generator, or some sort of electronic power supply.

The *load* is any device or circuit powered by electricity. It can be as simple as a light bulb or as complex as a modern high-speed computer.

The electricity provided by the source has two basic characteristics, called *voltage* and *current*. These are defined as follows:

Voltage

The electrical "pressure" causes free electrons to travel through an electrical circuit, also known as electromotive force (EMF).

There are two principal types of source, namely voltage source and current source. Sources can be either independent or dependent upon some other quantities. An independent voltage source maintains a voltage which is not affected by any other quantity. Similarly an independent current source maintains a current which is unaffected by any other quantity (Figure 7-9).

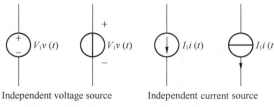

Figure 7-9 Symbols for independent sources

Some voltage (current) sources have their voltage (current) values varying with some other variables. They are called dependent voltage (current) sources or controlled voltage (current) sources (Figure 7-10).

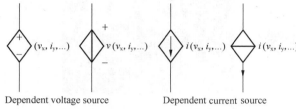

Figure 7-10 Symbols for dependent sources

Current

A collection of devices such as resistors and sources in which terminals are connected together by connecting wires is called an *electric circuit*. These wires converge in nodes, and the devices are called *branches* of the circuit (Figure 7-11).

The general circuit problem is to find all currents and voltages in the branches of the circuit when the intensities of the sources are known. Such a problem is usually referred to as circuit analysis.

Resistance

When voltage is applied to a piece of metal wire (Figure 7-12), the current I flowing through the wire is proportional to the voltage V across two points in the wire. This property is known as Ohm's law, which reads: $V = IR$ or $I + CV$. where R is called resistance, G is called conductance. The resistance R and the conductance G of the same piece of wire is related by $R = 1/G$. Resistance is measured in ohms (Ω) and conductance in Siemens (S or Ω).

The relationship between voltage, current, and resistance in an electrical circuit is fundamental to the operation of any circuit or device. Verbally, the amount of current flowing through a circuit is directly proportional to the applied voltage and inversely proportional to the circuit resistance. By explicit definition, one volt of electrical pressure can push one ampere of current through one ohm of resistance.

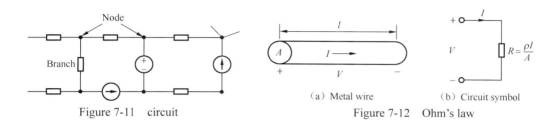

Figure 7-11　circuit　　　　　　　　Figure 7-12　Ohm's law

New Words

diagram ['daiəgræm]　n. 图
source [sɔ:s]　n. 电源
load [ləud]　n. 负载
terminal ['tə:minl]　n. 端子
designate ['dezigneit]　v. 指定
unbroken ['ʌn'brəukən]　a. 不断开的
push [puʃ]　v. 流出
arrow ['ærəu]　n. 箭头
generator ['dʒenəreitə]　n. 发电机
complex ['kɔmpleks]　a. 复杂的
electromotive [ilektrəu'məutiv]　a. 电动的
namely ['neimli]　ad. 即
independent [indi'pendənt]　a. 独立的
variable ['vεəriəbl]　n. 变量
converge [kən'və:dʒ]　v. 汇聚
node [nəud]　n. 节点
branch [brɑ:ntʃ]　n. 支路
intensity [in'tensiti]　n. 强度
proportional [prə'pɔ:ʃənl]　a. 成比例的
conductance [kən'dʌktəns]　n. 电导率
Siemens ['si:mənz]　n. 西门子
fundamental [ˌfʌndə'mentl]　a. 基本的
verbally ['və:bəli]　ad. 口头上
explicit [iks'plisit]　a. 明确的
volt [vəult]　n. 伏特
ampere ['æmpeə(r)]　n. 安培

Phrases and Expressions

block diagram　框图
as long as　只要
be known as　被称为
direct current (DC)　直流
in practice　在实际中
power supply　电源
electromotive force (EMF)　电动势
voltage source　电压源
current source　电流源
vary with　随着……的变化而变化
be proportional to　与……成比例
Ohm's law　欧姆定律
be directly proportional to　与……成正比
be inversely proportional to　与……成反比

EXERCISE 3

Choose the best electronic component according to the information given.

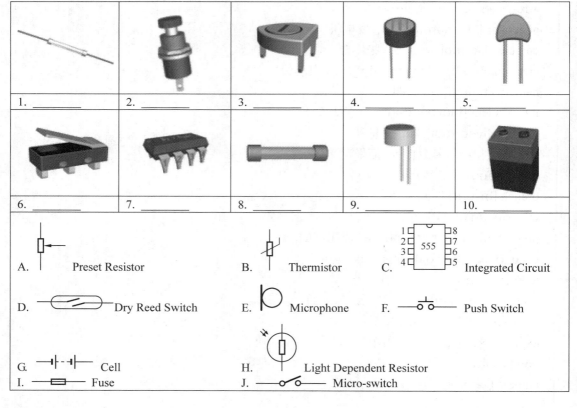

Unit 7 Electronic Components and Circuit

EXERCISE 4

Abbreviations are very useful in practical work. Read them and then translate them into corresponding Chinese terms.

1. PCM	Pulse Code Modulation	_____
2. P ds	Potential Differences	_____
3. PE	Pressure Element	_____
4. PDR	Power Direction Relay	_____
5. PF	Power Factor	_____
6. PL	Pipe Line	_____
7. PO	Power Output	_____
8. PT	Potential Transformer	_____

| Part II   Glance at Electronic Component Structures |

The following is the structure of transistor.

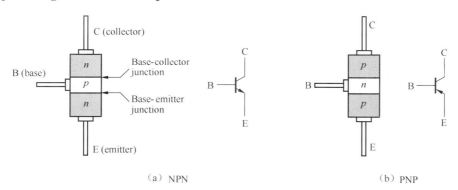

(a) NPN   (b) PNP

Explanations of the Transistor Terms

No.	Name	Explanation
1	Collector	集电极
2	Base	基极
3	Emitter	射极
4	Base-collector junction	基极集电极结
5	Base-emitter junction	基极射极结
6	NPN	NPN 型晶体管
7	PNP	PNP 型晶体管

EXERCISE 5

The following are some electronic components. You are required to choose the suitable words and phrases given below.

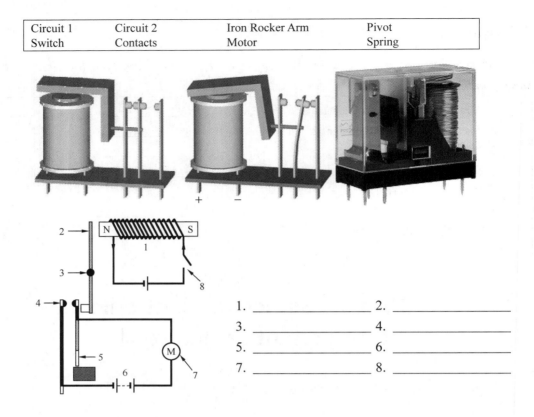

1. _____ 2. _____
3. _____ 4. _____
5. _____ 6. _____
7. _____ 8. _____

| Part III   Simulated Writing |

Section A   Match Your Skill

The following is the anatomy of light emitting diode (LED), and you can understand the name of Light Emitting Diode (LED).

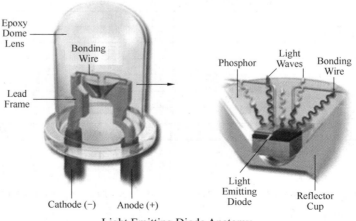

Light Emitting Diode Anatomy

Explanations of the Toll Bits

No.	Name	Explanation
1	Epoxy Dome Lens	环氧树脂封装的透光帽
2	Bonding Wire	焊线
3	Lead Frame	引线框
4	Cathode (−)	阴极
5	Anode (+)	阳极
6	Phosphor	荧光物质
7	Light Waves	光波
8	Light Emitting Diode	发光二极管
9	Reflector Cup	反光杯座

EXERCISE 6

Match the words or phraes on the left with their meanings on the right.

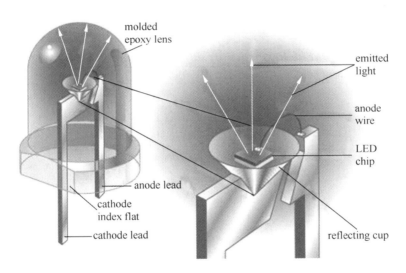

1. Molded epoxy lens
2. Anode lead
3. Cathode index flat
4. Cathode lead
5. Reflecting cup
6. LED chip
7. Anode wire
8. Emitted light

A. 阴极引线面
B. 反光杯
C. 阳极丝
D. 密封环氧树脂透镜
E. 发射光
F. 阳极引线
G. 阴极引线
H. 发光二极管芯片

Section B Have a Try

This section will help you to understand several forms of machining

Internal Operations

Boring — An operation in which a boring tool enters the work-piece axially and cuts along an internal surface to form different features. The boring tool is a single-point cutting tool, which can be set to cut the desired diameter by using an adjustable boring head. Boring is commonly performed after drilling a hole in order to enlarge the diameter or obtain more precise dimensions. On a turning machine, a variety of features can be formed, including steps, tapers, chamfers, and contours. These features are typically machined at a small radial depth of cut and multiple passes are made until the end diameter is reached. For a finish turning operation, the cutting feed is calculated based on the desired surface roughness and the tool nose radius.

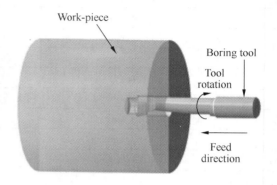

EXERCISE 7

This section is to test your ability to identify different operations.

Bored hole	Bored step	Boring (Milling machine)

(1) Boring

(2)

(3)

(4)

(1) Boring   (2) _____   (3) _____   (4) _____

| Part IV Broaden Your Horizon — Practical Activity |

How to Repair NEC P8000 Printer Main Board Problem

If the complaint for NEC P8000 printer (Figure 7-13) was a missing line across the printout, immediately we will suspect the printer head or the ribbon or even sometime the ribbon mask. Among so many brands of dot matrix printer, only this model have one common fault which is that the printer head and the printer driver ICs.

Figure 7-13

If you just replace the printer head without changing the main board or checking the printer driver ic for any short circuit, then chances are high that once you switch on the printer, the printer head would immediately breakdown again (Figure 7-14).

Thus it is wise for you, if you ever comes across this type of printer, besides replacing the printer head, you must check the six printer driver IC on the main board. As for the printer head, you can buy a new one or send to a specialist to refurbish it.

Now the real question is, how do we check or test the printer driver ICs? It's simple, remove the main board and you will observe that there are six ICs that have the same part number which is the STA476A. If one of them shorted, it will short the printer head pin. Each printer driver IC control 4 pins thus six printer IC would take care 24 pins (Figure7-15).

That's why in the market, they called it as 24 pin dot matrix printer.

Normally I would use huntron tracker (Figure 7-16) to compare the signature between a good and a bad IC. It is very easy to find the culprit when you can compare them.

Using an analog meter also can do the job. Compare the resistance of the entire printer driver ICs leg, if there are any shorted printer driver IC, the meter would register the reading and usually the needle would kick to most far right. Sometimes it could be more than one IC shorted. Once you have determined which IC that gave way, solder it out and replace with a new one and you have just completed a printer repair job.

Figure 7-14

Figure 7-15

Whether you are checking the NEC P8000 printer (Figure 7-17) or any brand of dot matrix printers, use the procedure above to test out all the printer driver IC before you switch on the printer. This not only saves your new printer head, it will save you money of not having to buy a new main board.

Figure 7-16

Figure 7-17

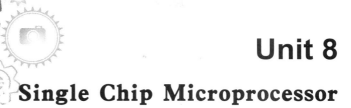

Unit 8
Single Chip Microprocessor

| Part I   Technical and Practical Reading |

Passage A   Introduction to Single Chip Microprocessor and Its Circuit

Single chip is a complete computer system integrated on a single chip. Even though most of its features in a small chip, it has a need to complete the majority of computer components: CPU, memory, internal and external bus system, most will have the core. At the same time, it also integrates communication interfaces, timers, real-time clock and other peripheral equipment. And now the most powerful single chip microcomputer system can even integrate voice, image, networking, input and output complex system on a single chip (Figure 8-1).

Figure 8-1   A Single Chip

A microcontroller is a single integrated circuit, commonly with the following features (Figure 8-2):

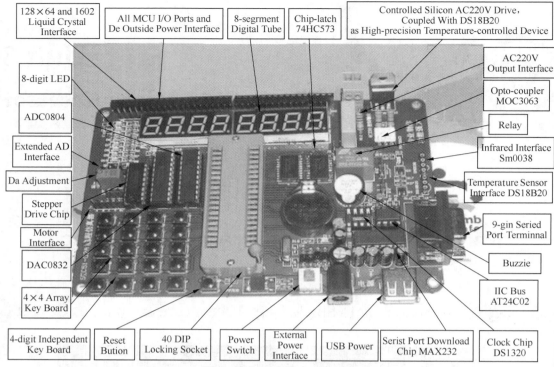

Figure 8-2 A Microcontroller

Central Processing Unit (CPU) — ranging from small and simple 4-bit processors to sophisticated 32- or 64-bit processors;

Input/Output interfaces such as serial ports (UARTs);

Other serial communication interfaces like Serial Peripheral Interface and Controller Area Network for system interconnection;

Peripherals such as timers and watchdog;

RAM for data storage;

ROM, EPROM, EEPROM or Flash memory for program storage;

Clock generator — often an oscillator for a quartz timing crystal, resonator or RC circuit;

Include analog-to-digital converters.

Microcontroller architectures are available from many different vendors in so many varieties. Among these are the chief 8051, Z80 and ARM derivatives. Most microcontrollers use a Harvard architecture: separate memory buses for instructions and data, allowing accesses to take place concurrently. In contrast to general-purpose CPUs, microcontrollers do not have an address bus or a data bus, because they integrate all the RAM and non-volatile memory on the same chip as the CPU. Because they need fewer pins, the chip can be placed in a much smaller, cheaper package (Figure 8-3).

The single chip microprocessor is also known as microcontroller because it was first used in the field of industrial control. The single chip microprocessor was developed from the dedicated processor only with CPU on the chip. The earliest design concept was to integrate a large number of peripherals and CPUs into a single chip so that the computer system could become smaller and more easily integrated into the complex and demanding control devices on their volume. The Z80 in INTEL was one of the first processors in accordance with the design idea. From then on, the development of SCM and dedicated processor parted away.

The single chip microprocessor is more suitable for embedded systems than the dedicated processor, so it is more widely used. It is a type of microprocessor emphasizing self-sufficiency (no connection to hardware) and cost-effectiveness in contrast to a general-purpose microprocessor used in a PC. Almost every piece of electronic and mechanical product used in modern human life will have a single-chip microprocessor. Cell-phone, telephone, calculator, home appliances, electronic toys, handheld computers and computer accessories such as a mouse are equipped with 1-2 single chips. And personal computers also have a large number of single-chip microcomputers that work. The number of SCMs is not only far exceeds the number of PCs, but the number of other integrated computing devices as well, even more than the number of human beings.

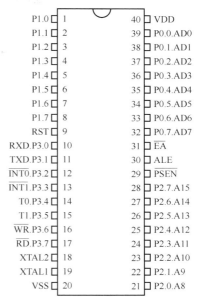

Figure 8-3   40-fin DIP

Notes

1. Most microcontrollers can use a Harvard architecture: separate memory buses for instructions and data, allowing accesses to take place concurrently.本句为简单句。由 allowing...所引导的结构，在句中做伴随状语。本句可译为：多数单片机使用哈佛结构：用于指令和数据的存储总线分离，可使存储同时进行。

2. Vehicles are generally equipped with more than 40 single-chips, and complex industrial control systems may even have hundreds of single-chips that do work at the same time 本句可译为：汽车一般配备 40 多台单片机，复杂的工业控制系统甚至可能有数百台单片机在同时起作用。

New Words

integrate ['intigreit]   v. 集成
memory ['meməri]   n. 内存
internal [in'tə:nl]   a. 内部的
external [eks'tə:nl]   a. 外部的
bus [bʌs]   n. 总线

timer ['taimə]   n. 定时器
real-time ['ri:əltaim]   a. 实时的
peripheral [pə'rifərəl]   a. 外围的
                          n. 外设
networking ['netwə:kiŋ]   n. 网络化
microcontroller [maikrəukən'trəulə]   n. 微控制器
processor ['prəusesə]   n. 处理器
serial ['siəriəl]   a. 串行的
interconnect [,intə(:)kə'nekt]   n. 相互连接
watchdog ['wɔtʃdɔg]   n. 看门狗
quartz [kwɔ:ts]   n. 石英
crystal ['kristl]   n. 晶体
resonator ['rezəneitə]   n. 谐振器
analog-to-digital ['ænəlɔgtu'didʒitl]   a. 模数的
converter [kən'və:tə(r)]   n. 转换器
opto-coupler [,ɔptəu'kʌplə]   n. 光电耦合器
sensor ['sensə]   n. 传感器
buzzle [bʌzl]   n. 蜂鸣器
flip-latch [fliplætʃ]   n. 锁存器
stepper ['stepə]   n. 步进电机
segment ['segmənt]   n. 段
architecture ['ɑ:kitektʃə]   n. 结构
derivative [di'rivətiv]   n. 衍生品
access ['ækses]   n. 存储，进入
general-purpose ['dʒenərəl'pə:pəs]   a. 通用的
non-volatile ['nɔn'vɔlətail]   a. 永久性的
pin [pin]   n. 管脚
embed [im'bed]   v. 嵌入
self-sufficiency [,selfsə'fiʃənsi]   n. 自供应（不用外接硬件）
department [di'pɑ:tmənt]   n. 部
exceed [ik'si:d]   v. 超过
compute [kəm'pju:t]   v. 计算

Phrases and Expressions

Single Chip Microprocessor (SCM)   单片机
real-time clock   实时时钟
peripheral equipment   外设
Central Processing Unit (CPU)   中央处理器

serial port 串行口
Universal Asynchronous Receiver/ Transmitter (UART) 异步串行接口
Random Access Memory (RAM) 随机存储器
Read Only Memory (ROM) 只读存储器
Erasable Programmable Read Only Memory (EPROM) 可擦可编程序只读存储器
Electrically Erasable Programmable Read Only Memory (EEPROM) 电可擦可编程序只读存储器
flash memory 闪存
clock generator 时钟发生器
analog-to-digital converter (ADC) 模数转换器
Alternating Current (AC) 交流
Universal Serial Bus (USB) 通用串行总线
Harvard architecture 哈佛结构
address bus 地址总线
data bus 数据总线
non-volatile memory 永久性存储器
Personal Computer (PC) 个人计算机
home appliance 家用电器

EXERCISE 1

The following is set of symbols denoting Working Safety. Choose the best symbol according to the information given below.

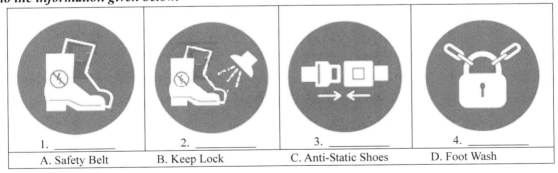

1. _____   2. _____   3. _____   4. _____
A. Safety Belt   B. Keep Lock   C. Anti-Static Shoes   D. Foot Wash

EXERCISE 2

Translate the following phrases into Chinese or English.

1. flash memory _____   2. _____ 单片机
3. address bus _____   4. _____ 耐磨性
5. Light Emitting Diode _____   6. _____ 时钟发生器
7. Universal Serial Bus _____   8. _____ 模数转换器

Passage B Introduction to MCS51 series Single Chip Microprocessor

The early single-chip is 8-bit or all of the four. One of the most successful is INTEL's 8031, because the performance of a simple and reliable access to a lot of good praise. A single-chip microcomputer system MCS51 series has been developed since then in 8031. Based on single-chip microcomputer system of the system is still widely used until now. As the field of industrial control requirements increase in the beginning of a 16-bit single-chip, it is not ideal because the price has not been very widely used. After the 90's with the big consumer electronics product development, single-chip technology is a huge improvement. INTEL i960 Series with subsequent ARM in particular, a broad range of applications, quickly replaced by 32-bit single-chip 16-bit single-chip high-end status, and enter the mainstream market. Traditional 8-bit single-chip performance has been the rapid increase in processing power compared to the 80's to raise a few hundred times. At present, the high-end 32-bit single-chip frequency over 300MHz, the performance of the mid-90's close on the heels of a special processor, while the ordinary price of the model dropped to one U.S. dollars, the most high-end models, only 10 U.S. dollars. Contemporary single-chip microcomputer system is no longer only the bare-metal environment in the development and use of a large number of dedicated embedded operating system is widely used in the full range of single-chip microcomputer (Figure 8-4).

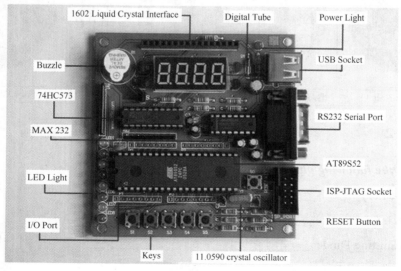

Figure 8-4

Unit 8 Single Chip Microprocessor

Now take an example for MCS51 series. AT89C2051 MCU, which is a series of 51 members, is the 8051 version of SCM. Internal comes with a programmable EPROM 2 k bytes of high-performance microcontrollers. With the industry standard MCS-51 orders and pin-compatible, so it is a powerful micro-controller, many embedded control applications, it provides a highly flexible and effective solutions. AT89C2051 has the following characteristics: 2 k bytes EPROM, 128 bytes RAM, 15 I / O lines, two 16-bit timers / counters, five vector two interrupt structure, a full two-way serial port, and includes precision analog comparator and on-chip oscillator, a 4.25 V to 5.5 V voltage scope of work and 12 MHz/24MHz frequency, and also offers the encryption array of two program memory locking, power-down and the clock circuit. In addition, AT89C2051 also supports two kinds of software-selectable power-saving mode power supply. During my free time, CPU stop and let RAM, timer / counter, serial port and interrupt system continue to work. Power-down can preserve the contents of RAM, but will stop oscillator and prohibit all the other functions until the next hardware reset.

MCU clock signal is used to provide various micro-chip microcontroller operation of the benchmark time, and the clock signal is usually used by the form of two circuits: the internal and external oscillation. MCS-51 has a microcontroller internal oscillator for a reverse of the high-gain amplifier, pin XTAL1 and XTAL2 are here to enlarge the electrical inputs and outputs, as in-house approach, a simple circuit, from the clock signal relatively stable, and actually used often in this way, as shown in Figure 8-5 in its external crystal oscillator (crystal) or ceramic resonator constituted an internal oscillation, an on-chip high-gain reverse amplifier and a feedback component of the chip quartz crystal or ceramic resonator together to form a self oscillator and generate oscillation clock pulse. Figure 8-5 in the external crystal and capacitors C1 and C2 constitute a parallel resonant circuits, their stability from the oscillation frequency, rapid start-up role, and its value are about 30pF, crystal frequency of elections 12 MHz.

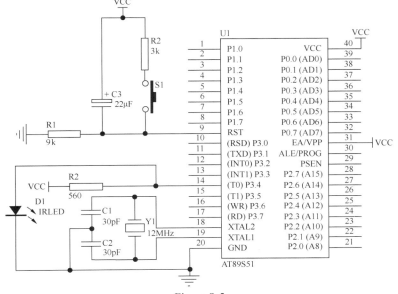

Figure 8-5

In order to initialize the internal MCU some special function register to be reset by the way, will reset after the CPU and system components identified in the initial state, and from the initial state began work properly. MCU is reset on the circuit to achieve, in the normal operation of circumstances, as long as the RST-pin on a two machine cycle time over the high, can cause system reset, but if sustained for the RST-pin HIGH, in a circle on the MCU reset state. After the system will reset input / output (I / O) home port register for the FFH, stack pointer SP home for 07 H, SBUF built-in value for the indefinite, all the rest of the register for 0, the status of internal RAM from the impact of reduction, On the system, when the contents of RAM is volatile. Reset operation uses two situations in which a power-on reset and manual (switch) reduction. The system uses a power-on reset mode. Figure 8-5 in the R0 and C0 formed a power-on reset circuit and its value for R for 4.7kΩ, C for the 10μF.

New Words

bit [bit]　　n.　位
performance [pəˈfɔːməns]　n.　性能
reliable [riˈlaiəbl]　a.　可靠的
high-end [hai] [end]　a.　高端的
status [ˈsteitəs]　n.　地位
mainstream [ˈmeɪnstriːm]　n.　主流
bare-metal [bɛəˈmetl]　a.　裸机的
socket [ˈsɔkit]　n.　插座
byte [bait]　n.　字节
compatible [kəmˈpætəbl]　a.　兼容的
counter [ˈkauntə]　n.　计数器
vector [ˈvektə]　n.　向量
comparator [ˈkɔmpəreitə]　n.　比较器
on-chip [ɔntʃip]　a.　片内的
oscillator [ˈɔsileitə]　n.　振荡器
encryption [inˈkripʃən]　n.　加密
array [əˈrei]　n.　阵列
locking [ˈlɔkiŋ]　n.　加锁
power-down [ˈpauədaun]　n.　掉电
preserve [priˈzəːv]　v.　保存
reset [ˈriːset]　n.　复位
benchmark [bentʃmɑːk]　n.　基准
in-house [inˈhaus]　a.　内部的
ceramic [siˈræmik]　n.　陶瓷
stability [stəˈbiliti]　n.　稳定性

Unit 8 Single Chip Microprocessor

start-up [stɑ:tʌp]   n. 起振
initialize [i'niʃəlaiz]   v. 初始化
register ['redʒistə]   n. 寄存器
built-in ['bilt'in]   a. 内置的
volatile ['vɔlətail]   a. 不定的

Phrases and Expressions

mainstream market   主流市场
processing power   处理能力
special processor   专用处理器
operating system   操作系统
Personal Digital Assistant (PDA)   掌上电脑，个人数字助理
crystal oscillator   晶体振荡器
ceramic resonator   陶瓷谐振器
self oscillator   自激振荡器
parallel resonant circuit   并联谐振电路
stack pointer   堆栈指针

EXERCISE 3

Match the term with the picture given.

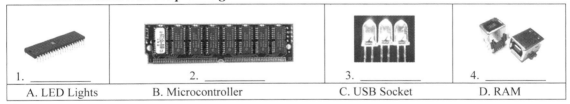

1. _____	2. _____	3. _____	4. _____
A. LED Lights	B. Microcontroller	C. USB Socket	D. RAM

EXERCISE 4

Abbreviations are very useful in practical work. Read them and then translate them into corresponding Chinese terms.

1. QC	*Quality Control*	_____
2. R.B.	*Roller Bearing*	_____
3. R.C.	*Reaction Coupling*	_____
4. S.E.	Standard Error	_____
5. S.H.P.	Shaft Horse-power	_____
6. Sh. S.	*Sheet Steel*	_____
7. S/M	*Surface-to-mass Ratio*	_____
8. SL	*Square Law*	_____

Part II Glance at Single Chip Microprocessor Structures

The following is the structure of single chip microprocessor.
Sample

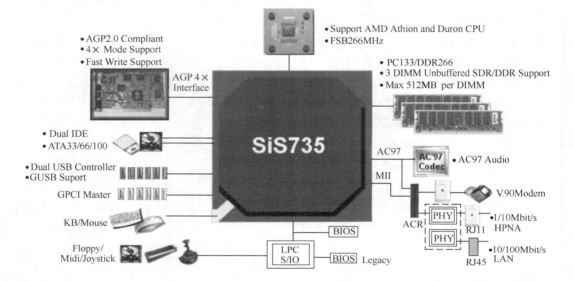

Explanations of the Single Chip Microprocessor Terms

No.	Name	Explanation
1	AGP 2.0 Compliant	可兼容的加速图形接口 2.0
2	4×Mode Support	支持 4×模式
3	Fast Write Support	支持快写模式
4	AGP 4×Interface	加速图形 4×接口
5	Dual IDE	一对电子集成驱动器
6	ATA33/66/100	微型计算机系统接口 33/66/100 系列
7	Dual USB Controller	一对通用串行总线控制器
8	G USB Support	支持千兆通用串行总线
9	GPCI Master	千兆主机外设连接接口
10	KB/Mouse	键盘/鼠标
11	Floppy/Midi/Joystick	软盘/音乐设备数字接口/摇杆
12	LPC S/IO	本地过程调用机构的 S/IO 系统
13	BIOS	基本输入输出系统
14	BIOS Legacy	基本输入输出系统第二代
15	10/100Mbit/s LAN	10/100 兆位局域网
16	PHY	实体层
17	ACR	存取控制寄存器
18	1/10Mb HPNA	1/10 兆位家庭电话线网络适配器
19	V.90 Modem	V.90 调制解调器

Unit 8 Single Chip Microprocessor

续表

No.	Name	Explanation
20	MII	介质无关接口
21	AC97 Audio	AC97 音频
22	Max. 512Mb per DIMM	双列直插式内存模块，每个最大为 512 兆位
23	3 DIMM Unbuffered SDR"/"DDR Support	3 个双列直插式内存模块，支持无缓冲的同步动态随机存储器/双倍速率同步动态随机存储器
24	PC133"/"DDR266	个人计算机速率 133MHz/双倍速率同步动态随机存储器速率 266 MHz
25	FSB266MHz	前端总线频率 266MHz
26	Support AMD Athlon and Duron CPU	支持 AMD 公司的速龙系列和杜龙系列 CPU

EXERCISE 5

The following is a single chip microprocessor. You are required to choose the suitable term for each part.

| CPU | Expansion board | Battery | AGP slot | L2 Cache |
| PCI slots | Keyboard connector | ISA slots | RAM (SIMM modules) | |

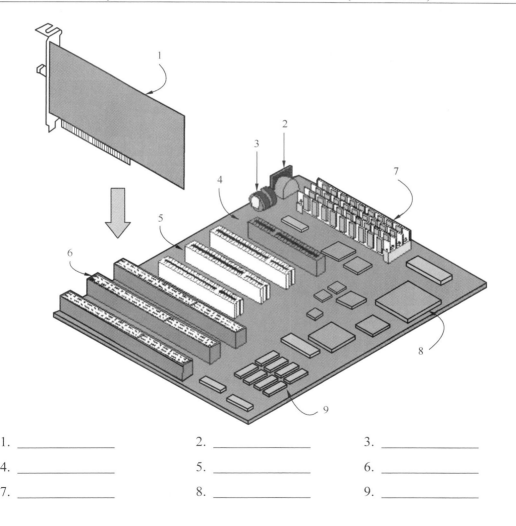

1. _____
2. _____
3. _____
4. _____
5. _____
6. _____
7. _____
8. _____
9. _____

Part III Simulated Writing

Section A Match Your Skill

The following is a single chip microprocessor, and you can understand its name.

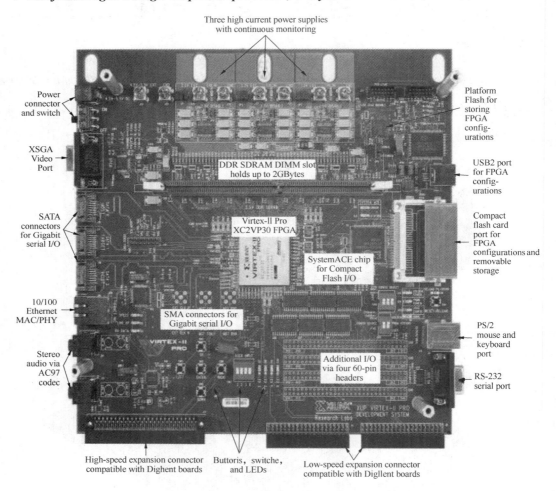

Explanations of the Machine Tool Terms

No.	Name	Explanation
1	Platform Flash for storing FPGA configurations	现场可编程门阵列板式闪存
2	USB2 port for FPGA configurations	用于现场可编程门阵列存储器的系列 2 通用串行总线口
3	Compact flash card port for FPGA configurations and removable storage	用于现场可编程门阵列存储器的闪存卡口和可移动存储装置
4	PS/2 mouse and keyboard port	PS/2 系列鼠标和键盘口
5	RS-232 serial port	RS-232 系列串行口

Unit 8 Single Chip Microprocessor

续表

No.	Name	Explanation
6	Power connector and switch	电源接口和开关
7	XSGA video port	高分辨率视频口
8	SATA connectors for Gigabit serial I/O	用于吉位串行输入/输出口的串行 ATA 连接器
9	10/100 Ethernet MAC/PHY	10/100 系列以太网媒体存取控制/实体层
10	Stereo audio AC97 codec	AC97 系列立体声音频编解码器
11	Three high current power supplies with continuous monitoring	三个可进行连续监控的高电流电源
12	High-speed expansion connector compatible with Digilent boards	可与 Digilent 系列板兼容的高速扩展接头
13	Buttons, switches, and LEDs	按钮、开关和发光二极管
14	Low-speed expansion connector compatible with Digilent boards	可与 Digilent 系列板兼容的低速扩展接头
15	DDR SDRAM DIMM slot holds up to 2GBytes	容量可达 2GB 的双倍速率同步动态随机存储器插槽/双列直插式内存模块插槽
16	XC2VP30 FPGA	XC2VP30 系列现场可编程门阵列存储器
17	SMA connectors for Gigabit serial I/O	用于吉位串行输入/输出口的系统管理连接器
18	System ACE chip for compact flash I/O	提供闪存卡输入/输出口的 ACE 系统芯片
19	Additional I/O via four 60-pin headers	有 4 个 60 针的附加输入/输出口

EXERCISE 6

Complete the information by translating the parts given in Chinese.

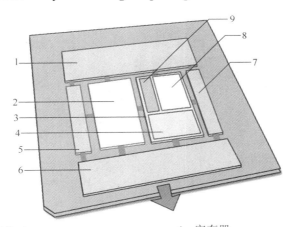

1. L2 Cache
2. Control unit
3. Instruct unit
4. Register
5. L1 Cache (instructions)
6. Input – output management unit
7. L1 Cache (data)
8. FPU
9. ALU (Analog Lines Unit)

A. 寄存器
B. 指令单元
C. 第二层快取存储器
D. 输入—输出管理单元
E. 控制单元
F. 浮点运算单元
G. 模拟线路单元
H. 第一层快取存储器（指令）
I. 第一层快取存储器（数据）

Section B Have a Try

This section will help you to understand several forms of machining.

Internal Operations

Reaming — An operation in which a reamer enters the work-piece axially and enlarges an existing hole to the diameter of the tool. Reaming removes a minimal amount of material and is often performed after drilling to obtain both a more accurate diameter and a smoother internal finish. A finish reaming operation will use a slower cutting feed to provide an even better finish.

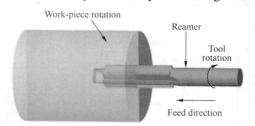

EXERCISE 7

This section is to test your ability to identify different operations.

| Bored taper | Reamed hole | Reaming (Milling machine) |

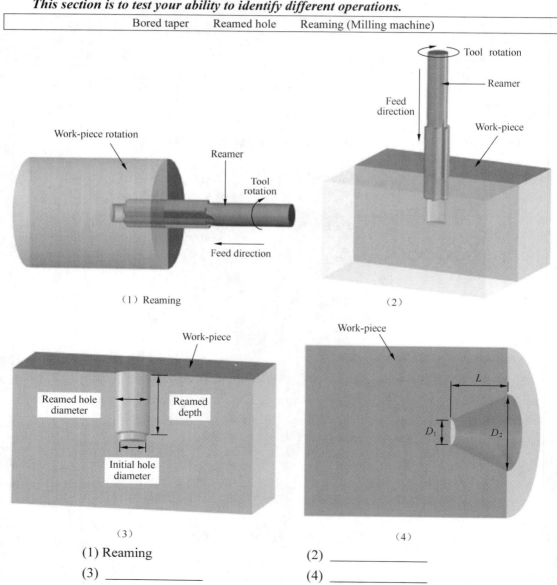

(1) Reaming (2)
(3) (4)

(1) Reaming
(3) _____

(2) _____
(4) _____

| Part IV   Broaden Your Horizon — Practical Activity |

The 3D LED Cube Completed

This part shows some stuff done over the past month. Now, on to the work completed.

The first thing we did was a lot of programming. I wondered if we could time modulate the bicolor LEDs to get 16 levels each of red and green, giving a total of 256 colors. To do this well, it would take a double-buffered interrupt driven display, so the first thing I did was to get this running, with a simple rule: color 0 is 0 slice out of 15, color 1 is 1 slice out of 15, up to color 15 is all 15 out of 15. We estimated we'd need about 50 frames per second to avoid visual flickering, and each interrupt draws 8 pixels, so to get all 64 LEDs updating 50 times a second with a pattern cycle length of 15 would take $50*(64/8)*15 = 6000$ interrupts a second. Since our PIC runs 8 MIPS, we'd have $2^{23}/6000 = 1400$ cycles per interrupt. After programming this, there was still weird flicker on some colors, and it took a lot of work to eliminate this.

We also added 4 buttons for control and debugging purposes. Here is the resulting circuit (Figure 8-6 and Figure 8-7).

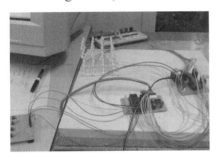
Figure 8-6

Figure 8-7

Here are the buttons to control the cube (Figure 8-8).

Here is the main CPU board, with plugs to run the timing board, connect to the button board, and a switch to toggle programming/running modes (Figure 8-9).

Figure 8-8

Figure 8-9

It shows the bottom of the board (Figure 8-10).

Figure 8-11 shows the board responsible for taking clock ticks and control signals and lighting the LEDs.

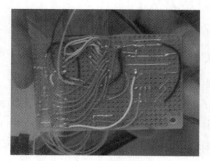
Figure 8-10

Figure 8-11

Wanting to try some other spacing before making 8×8×8 cubes, we created a smaller one. First a bag of LEDs (Figure 8-12).

Cubes were soldered into a smaller template in Figure 8-13.

Figure 8-12

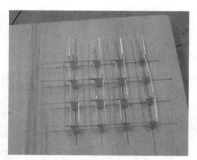
Figure 8-13

Figure 8-14 shows resulting in the Mini-Cube.

We made an Uber-template and some of our wire, along with a wiring design we concocted, and created a bigger 4×4×4 cube. Two benefits this design has are all wires leaving the base instead of some on the sides, and it is widely-spaced since we used some stiff wire instead of LED to lead for connection. The downside is that it was much harder to solder this together, a problem needing to be solved before attempting the 8×8×8 cubes (Figure 8-15).

Figure 8-14

The big one had a new wiring mode letting all wires leave the bottom, making it look much better. Minor code changed and it was up and running (Figure 8-16).

We moved some of the PIC ports around to allow 4 debugging lights (bicolor one port output), nice control buttons, etc., all on one board (Figure 8-17).

Unit 8 Single Chip Microprocessor

Figure 8-15

Figure 8-16

Figure 8-18 shows the board top.

Figure 8-17

Figure 8-18

We were also experimenting with a single chip solution, shown (Figure 8-19) here from the top (this replaces everything except the power and PIC).

And the insanely wired bottom in Figure 8-20:

Figure 8-19

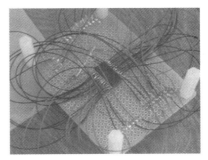

Figure 8-20

The LED Cube — Hardware Update

We had some power problems — it appeared the circuit drawed more amps than multimeters report, since multimeters seemed to time average the amounts, and we had spikes. After some issues, this seems solved.

We are now in the process of making PCBs for 4×4×4 standalone, 4×4×4 with programmer, and the final 8×8×8 standalone PCB.

Unit 9
Introduction to Motors

| Part I Technical and Practical Reading |

Passage A The Motor Basics (I)

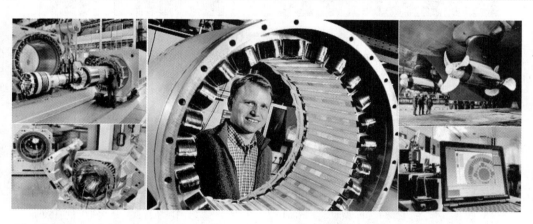

The function of the motor is to convert electrical energy to mechanical energy. Thus, the motor is also classified as an electromechanical device/machine.

Basics of Electromagnetic Forces

Electromagnetic Force

The direction of magnetic flux produced by a permanent magnet is always from N-pole to S-pole.

When a conductor is placed in a magnetic field and current flows in the conductor, the magnetic field and the current interact with each other to produce force. The force is called "Electromagnetic force" (Figure 9-1).

The Fleming's left hand rule determines the direction of the current, the magnetic force and the flux. Stretch the thumb, the index finger and the middle finger of your left hand as shown in Figure 9-2.

Unit 9 Introduction to Motors

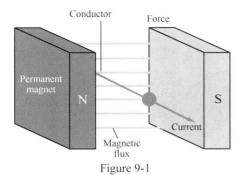

Figure 9-1

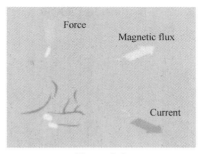

Figure 9-2

When the middle finger is the current and the index finger the magnetic flux, the direction of the force is given by the thumb.

Magnet field produced by current

The magnetic fields produced by the current and the permanent magnets work to produce electromagnetic force.

When the current flows in the conductor toward the reader, the magnetic field in the CCW direction will be produced around the current flow by the right-handed screw rule (Figure 9-3).

Interference of a line of magnetic force

The magnetic fields produced by the current and the permanent magnets interfere each other (Figure 9-4).

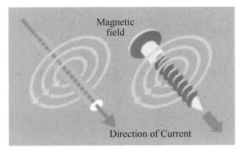

Figure 9-3

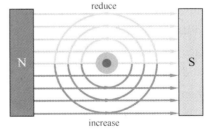

Figure 9-4

The line of magnetic force distributed in the same direction acts to increase its strength, while the flux distributed in the opposite direction acts to reduce its strength.

Electromagnetic force production

The line of magnetic force has a nature to return to the straight line by its tension like an elastic band.

Thus, the conductor is forced to move from where the magnetic force is stronger to where it is weaker (Figure 9-5).

Torque production

Electromagnetic force is obtained from the equation:

F(force)=B(magnetic flux density) · I(current) · L(length of conductor)

Figure 9-6 illustrates the torque obtained when a single-turn conductor is placed in the magnetic field.

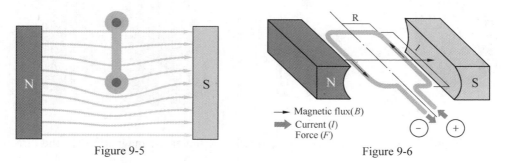

Figure 9-5   Figure 9-6

The torque produced by the single conductor is obtained from the equation:

$T'=F \cdot R$

- T' (torque)
- F (force)
- R (distance from the center to conductor)

Here, there are two conductors present:

$T=2 \cdot T'=2 \cdot F \cdot R$

Principles of Motor Motion

1. Place Permanent Magnet

Place permanent magnets so that different polarities face each other.

Thus, a parallel magnetic field will be created (Figure 9-7).

2. Place Armature

Place the armature in the parallel magnetic field and conduct magnetization to have N-polarity at ① and S-polarity at ② and ③.

Then, the armature will start CW rotation due to magnetic attraction and repulsion (Figure 9-8).

 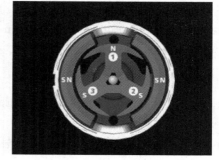

Figure 9-7   Figure 9-8

3. Magnetize Armature

Wind the coil and flow the current so that the armature will be magnetized (Figure 9-9).

4. To Keep Rotation

To keep the armature running, the current direction should be changed so that the N-polarity stays at upper portion and S- polarity stays at under portion of the armature (Figure 9-10).

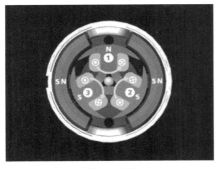
Figure 9-9

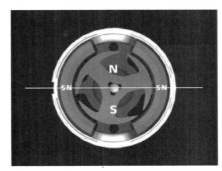
Figure 9-10

5. Switch Current Direction

To change the current direction, coil shall be connected to the metal strips (commutator) shaped as a cylinder divided into three (Figure 9-11).

6. Place Brush

To keep CW rotation, apply plus (+) voltage to the left and minus (−) to the right brush.

If CCW rotation is required, apply minus (−) voltage to the left and plus (+)to the right brush, or reverse the polarity of the magnet (Figure 9-12).

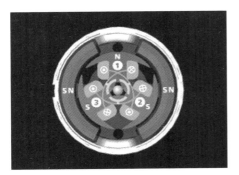
Figure 9-11

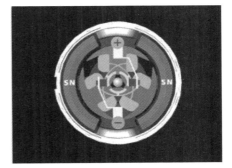
Figure 9-12

Notes

1. Place the armature in the parallel magnetic field and conduct magnetization to have N-polarity at ① and S-polarity at ② and ③ .全句为祈使句；Place 和 conduct 为并列结构。全句译为：在磁场的平行方向置入电枢，同时产生了磁化现象，①是 N 极、②和③是 S 极。

2. To keep the armature running, the current direction should be changed so that the N-polarity stays at upper portion and S- polarity stays at under portion of the armature.本句是目的状语从句。To keep the armature running 是不定式用作目的状语，置于居首为强调。本句译为：为了保持电枢旋转，应改变电流方向，使 N 极总在电枢的上部，S 极总在电枢的下部。

New Words

convert [kənˈvəːt]　v. 转换
electromechanical [iˌlektrəllmiˈkænikəl]　a. 机电的
electromagnetic [ilektrəuˈmægnitik]　a. 电磁的
flux [flʌks]　n. 通量，磁通
magnet [ˈmægnit]　n. 磁铁
pole [pəul]　n. 极
interact [ˌintərˈækt]　v. 相互作用
clockwise [ˈklɔkwaiz]　ad. 顺时针
counterclockwise [ˌkauntəˈklɔkwaiz]　ad. 逆时针
interference [ˌintəˈfiərəns]　n. 干涉
distribute [disˈtribju(ː)t]　v. 分布
strength [streŋθ]　n. 强度
nature [ˈneitʃə]　n. 特性
equation [iˈkweiʃən]　n. 方程
illustrate [ˈiləstreit]　v. 图解
single-turn [ˈsiŋgltəːn]　a. 单圈的
principle [ˈprinsəpl]　n. 原理
motion [ˈməuʃən]　n. 运动
polarity [pəuˈlæriti]　n. 极性
armature [ˈɑːmətjuə]　n. 电枢
magnetize [ˈmægnitaiz]　v. 磁化
attraction [əˈtrækʃən]　n. 吸引
repulsion [riˈpʌlʃən]　n. 排斥
upper [ˈʌpə]　a. 上部的
portion [ˈpɔːʃən]　n. 部分
switch [switʃ]　v. 切换
commutator [ˈkɔmjuteitə]　n. 换向器
brush [brʌʃ]　n. 电刷
plus [plʌs]　a. 正号；正极；正数
minus [ˈmainəs]　a. 负的

Phrases and Expressions

magnetic flux　磁通
Fleming's left hand rule　弗莱明左手规则
right-handed screw rule　右手螺旋定则

line of magnetic force   磁力线
straight line   直线
divide into   分成
elastic band   松紧带

EXERCISE 1

The following is set of symbols denoting Working Safety. Choose the best symbol according to the information given below.

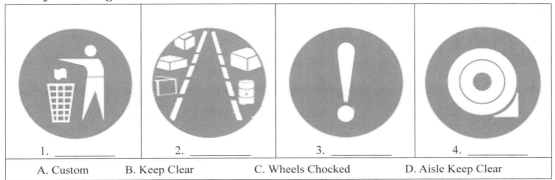

1. _____   2. _____   3. _____   4. _____

A. Custom   B. Keep Clear   C. Wheels Chocked   D. Aisle Keep Clear

EXERCISE 2

Translate the following phrases into Chinese or English.

1. magnetic flux   _____
2. _____   电磁力
3. permanent magnet   _____
4. _____   右手螺旋定则
5. magnetic field   _____
6. _____   磁力线
7. elastic band   _____
8. _____   直线

Passage B   The Motor Basics (II)

Function of Main Motor Parts

1. Produce Magnet Field

Black-colored portions are permanent magnets that maintain magnetic force, and are fixed to a motor-housing. The motor-housing itself produces a magnetic circuit (a magnetic field) with a core (a green-colored portion) and the permanent magnets as a path for the magnetic force to flow (Figure 9-13).

2. Provide Current

Current enters the motor from one of the terminals (a portion sticks out forward).

It flows to a winding (an orange-colored portion) via brushes (a brown-colored portion) and a commutator, and goes out of the motor from the other terminal via the brushes and the commutator (Figure 9-14).

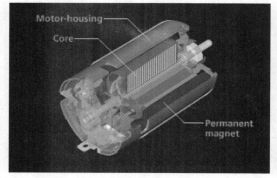
Figure 9-13

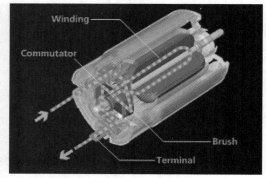
Figure 9-14

3. Output by Rotation

The current flows into the magnet field produced by the permanent magnets, and that produces electromagnetic force. The commutator switches the direction of the current in exact timing to have continuous rotation. The output is pulled out through a shaft (Figure 9-15).

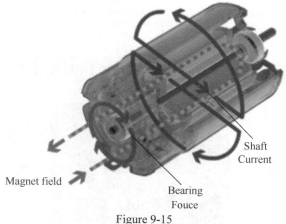
Figure 9-15

Explanation of Motor Parts (Figure 9-16 to Figure 9-20)

Figure 9-16　Completed motor

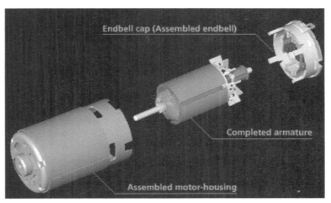

Figure 9-17　Exploded diagram of completed motor

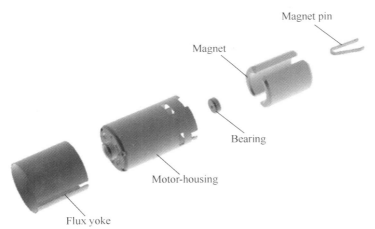

Figure 9-18　Exploded diagram of motor-housing

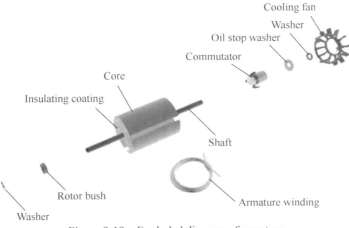

Figure 9-19　Exploded diagram of armature

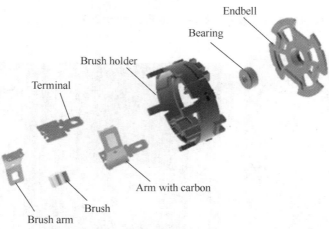

Figure 9-20　Exploded diagram of endbell cap

List of terms on motor performance

Term	Symbol	Unit
Input	P	W
Output	P	W
Maximum output	P max.	W
Voltage	V	V
Current	I	A
No-load current	I0	A
Stall current	Is	A
Efficiency	η	%
Maximum efficiency	η max.	%
Speed	N	r/min
No-load speed	N0	r/min
Torque	T	N·m, g·cm
Stall torque	Ts	N·m, g·cm

General Instructions for Use of Motors

In handling motors, be sure to read these general instructions to ensure safe, secure, and correct use of the motors.

Warning

1. Do not insert a lead or motor terminal into a home outlet. You might get an electric shock from this.

2. When applying current, do not touch a live part such as a current-carrying terminal. You might get an electric shock from this.

3. When applying current, keep your hand or finger off a rotary part. Otherwise, you might get hurt.

4. DON'T leave motor shaft locked while power is applied, as even a short-time lock-up may cause excess heat build up resulting in burning damage to the motor.

5. Depending on motor operating state (mounted state, load, and environmental temperature), the motor might generate excessive heat. Be careful not to burn yourself due to the heat.

Unit 9 Introduction to Motors

New Words

field [fiːld]　n. 场
fix [fiks]　v. 固定
motor-housing [ˈməutəˈhauziŋ]　n. 电机壳
rotation [rəuˈteiʃən]　n. 旋转
assemble [əˈsembl]　v. 组装
endbell [endbel]　n. 端承口
yoke [jəuk]　n. 轭
washer [ˈwɔʃə]　n. 垫圈
bush [buʃ]　n. 衬套
insulating [ˈinsjuleitiŋ]　a. 绝缘的
coating [ˈkəutiŋ]　n. 涂料
no-load [ˈnəuləud]　n. 空载
outlet [ˈautlet]　n. 插座
current-carrying [ˈkʌrəntˈkæriiŋ]　a. 载流的
lock-up [lɔkʌp]　n. 锁紧

Phrases and Expressions

magnet field　磁场
permanent magnet　永久磁铁
magnetic force　磁力
magnetic circuit　磁路
electromagnetic force　电磁力
endbell cap　端承口盖
magnet pin　磁性夹
insulating coating　绝缘涂料
cooling fan　冷却扇
brush holder　刷座
brush arm　刷臂
no-load current　空载电流
stall current　失速电流
no-load speed　空载转速
stall torque　失速转矩
electric shock　电击
live part　带电部件
build up　积聚

EXERCISE 3

Choose the best motor accessory according to the information given.

1. _____ 2. _____ 3. _____ 4. _____ 5. _____
A. Stator B. Rotor C. Spring D. Permanent Magnets E. Windings

EXERCISE 4

Abbreviations are very useful in practical work. Read them and then translate them into corresponding Chinese terms.

1. S.P.	Standard Pressure	_____	5. S.S.C.	Semi-steel Casting	_____
2. SR	Slant Range	_____	6. ST	Standard Temperature	_____
3. SS	Stainless Steel	_____	7. st. c.	Steel Casting	_____
4. ss	Spindle Speed	_____	8. str. st.	Structure Steel	_____

Part II Glance at Automatic Control Structures

The following is the structure of motor control.

Motor Control Overview

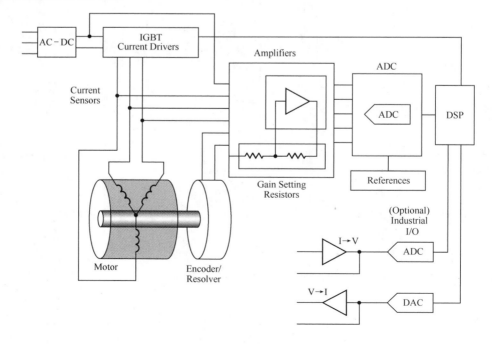

Unit 9 Introduction to Motors

Explanations of Terms

No.	Name	Explanation
1	AC—DC	交流—直流
2	IGBT Current Drivers	绝缘栅双极晶体管电流驱动器
3	Amplifers	放大器
4	ADC	模数转换器
5	Current Sensors	电流传感器
6	DSP	数字信号处理器
7	Gain Setting Resistors	增益设定电阻器
8	References	参考值
9	Industrial (Optional)	工业输入"/"输出口（可选）
10	Motor	电机
11	Encoder/Resolver	编码器/分解器
12	DAC	数模转换器

EXERCISE 5

Choose the suitable term given below.

Power supply	Terminal box	Bearing	Rotor
Drive pulley	Centrifugal switch	Cooling fan	Stator

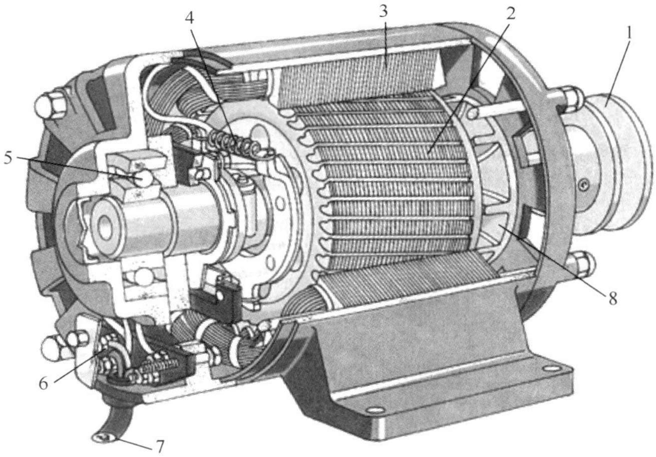

1. _____ 2. _____
3. _____ 4. _____
5. _____ 6. _____
7. _____ 8. _____

| Part III   Simulated Writing |

Section A   Match Your Skill

The following are some motor accessories, and you can learn their name.

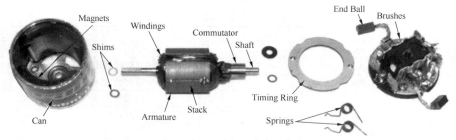

Explanations of the Motor Terms

No.	Name	Explanation
1	Magnets	磁铁
2	Can	罩
3	Shims	垫片
4	Windings	绕组
5	Armature	电枢
6	Stack	帧
7	Commutator	换向器
8	Shaft	传动轴
9	Timing Ring	定时环
10	Springs	弹簧
11	End Ball	端球
12	Brushes	电刷

EXERCISE 6

Match the words or phrases on the left with their meanings on the right.

Motor

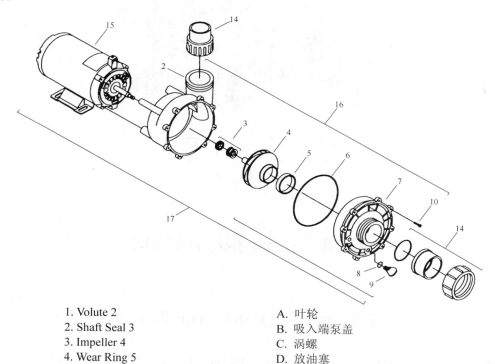

1. Volute 2
2. Shaft Seal 3
3. Impeller 4
4. Wear Ring 5

A. 叶轮
B. 吸入端泵盖
C. 涡螺
D. 放油塞

5. O-Ring 6
6. Suction Cover 7
7. Drain Plug 8
8. Drain Plug Adapter 9
9. Screw 10
10. Pump Union 14
11. Motor 15
12. Wet end complete 16
13. Pump complete 17

E. 轴封
F. 螺丝
G. 耐磨环
H. 电机
I. O 型环
J. 放油塞转接器
K. 液体部分总成
L. 泵总成
M. 泵组

Section B Have a Try

This section will help you to understand several forms of machining.

Internal Operations

Taping—An operation in which a tap tap enters the work-piece axially and cuts internal threads into an existing hole. The existing hole is typically drilled by the required tap drill size that will accommodate the desired tap. On a milling machine, the threads may be cut to a specified depth inside the hole (bottom tap) or the complete depth of a through hole (through tap).

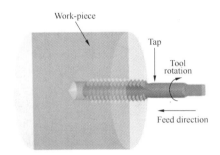

EXERCISE 7

This section is to test your ability to identify different operations.

Bored chamfer	Tapping (Milling machine)	Tapped hole

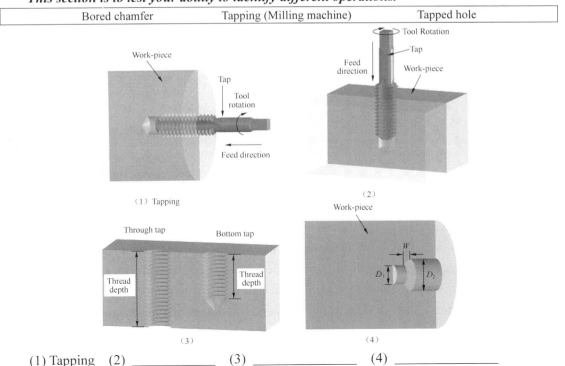

(1) Tapping   (2) _____   (3) _____   (4) _____

Part IV Broaden Your Horizon—Practical Activity

Mini Motor Repair

STEP 1: If you look at the end cap of the motor housing, you will see four metal corners bent at an angle to hold the end cap on. All you have to do is to take your pliers and bend them in the opposite direction so that you can remove the end cap. Then use a flathead screwdriver to pry the end cap off (Figure 9-21 to 9-24).

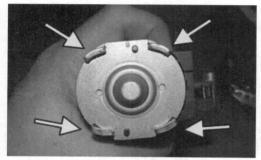

Figure 9-21

Figure 9-22

Figure 9-23

Figure 9-24

STEP 2: Pull the center shaft out from the motor case (Figure 9-25).

STEP 3: With the end cap removed and separated from the shaft, use your Digital Ohm Meter to test both outside terminals to see if there is any resistance. If you find that there is zero resistance at the terminals, then that would indicate that your circuit is closed when it should be open, this means your capacitor has a short in it. Locate the capacitor on the end cap. Use a pair of wire cutters to separate the connection, as shown in the figures. Be sure to position it so that it doesn't reconnect on accident. Use your Ohm Meter to retest both outside terminals again, and you should see that the circuit is now open, and indicates full resistance. This will confirm that you have corrected the problem with the motor (Figure 9-26 to 9-29).

Unit 9 Introduction to Motors

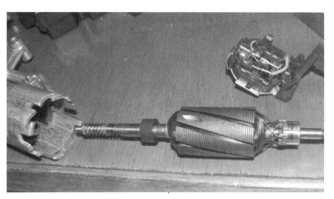

Figure 9-25

Figure 9-26

Figure 9-27

Figure 9-28

Figure 9-29

STEP 4: Now in order to put the shaft back into the motor case, you will need to reconnect the shaft back into the end cap. In order to do so, you will need to use your thin flat screwdriver to move the two spring loaded brushes out of the way, so that the end cap can fit back over the shaft (Figure 9-30 to 9-34).

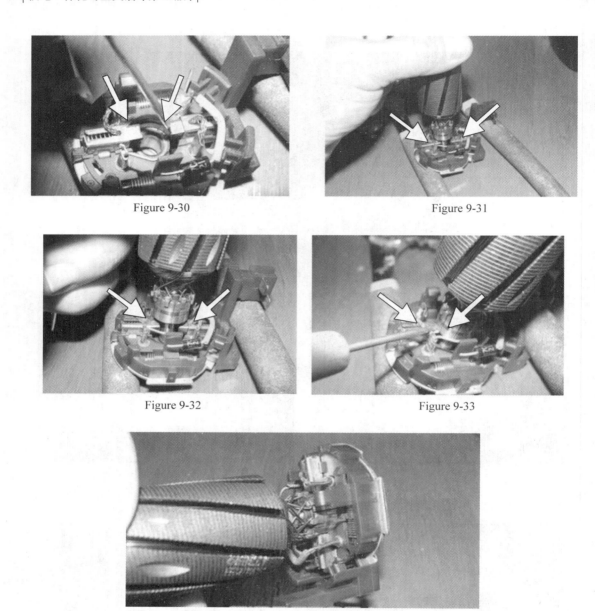

Figure 9-30

Figure 9-31

Figure 9-32

Figure 9-33

Figure 9-34

STEP 5: Now with both pieces connected, carefully hold them together while you slowly insert the shaft back into the motor case.

Tips: You should place something on the opposite side of the motor case to prevent the gear drive, from the shaft, from sliding through the motor case too quickly. The shaft and motor case are both highly magnetized and will have the tendency to pull the shaft quickly through the motor case. If you allow the shaft to slide too quickly through the motor case, your end cap will become separated from the shaft, and you will have to repeat this step over again.

Don't forget to seal the end cap properly, before putting the rest of the motor housing back together (Figure 9-35 to 9-36).

Unit 9 Introduction to Motors

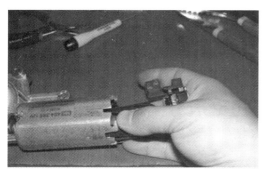

Figure 9-35

Figure 9-36

Unit 10
Introduction to Programmable Logic Controller

| Part I   Technical and Practical Reading |

Passage A   Programmable Logic Controller (PLC)

This section on programmable logic controllers illustrates just a small sample of their capabilities. As computers, PLCs can perform timing functions (for the equivalent of time-delay relays), drum sequencing, and other advanced functions with far greater accuracy and reliability than what is possible to use electromechanical logic devices. Most PLCs have the capacity for far more than six inputs and six outputs. Figure 10-1 shows several input and output modules of a single Allen-Bradley PLC.

With each module having sixteen "points" of either input or output, this PLC has the ability to monitor and control dozens of devices. Fit into a control cabinet, a PLC takes up little room, especially considering the equivalent space that would be needed by electromechanical relays to perform the same functions (Figure 10-2).

One advantage of PLCs that simply cannot be duplicated by electromechanical relays is remote monitoring and control via digital computer networks. Because a PLC is nothing more than a special-purpose digital computer, it has the ability to communicate with other computers rather easily. (Figure 10-3) shows a personal computer displaying a graphic image of a real liquid-level process (a pumping station for a municipal wastewater treatment system) controlled by a PLC. The actual pumping station is located miles away from the personal computer display.

Figure 10-1   PLC(programmable logic controller)

Figure 10-2   A control cabinet

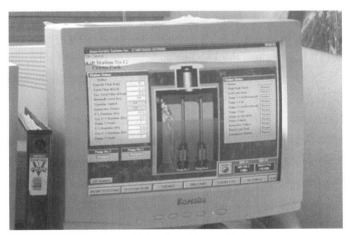

Figure 10-3   Remote monitoring

The characteristic structure of programmable logic controllers is illustrated in Figure 10-4 Provision is made to accept information from various input switching devices such as push-buttons, limit switches, and relay contacts. Also provided are terminals to which output devices such as solenoids, relay coils, and indicator lights can be connected. There are no direct-wired connections between the inputs and outputs. Instead, the switched conditions of the inputs are converted to logic-level signals that are inputs to a digital computer in the control unit. A program stored in the computer specifies which outputs should be energized on the basis of the present and past states of the input signals. Logic-level outputs from the control unit are converted to the voltage levels required to energize or de-energize the various output devices.

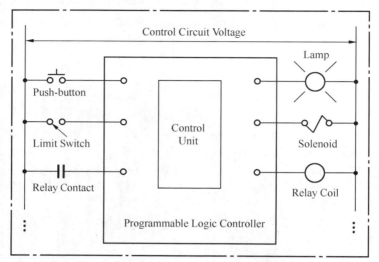

Figure 10-4 Characteristic structure of a programmable logic controller

The PLC software is comprised of the operating system, the basic software and the user program. The operating system contains all the statements and declarations for internal system functions. The basic software has a flexible interface to the operating basic functions. The basic software also contains some function blocks. The user program is the total of all the statements and declarations programmed by the user. The user program consists of a number of block types such as organization blocks (OBs), program blocks (PBs), function blocks (FBs), sequence blocks (SBs) and data blocks (DBs). They are available for structuring the user program and each of the blocks is used for different tasks.

Many control applications exist that require the switching (on or off) of various outputs as a function of the condition (on or off) of a number of input devices. This type of control is referred to as switching control of logic control. The relative simplicity of switching control makes it attractive for use in automatic machines and processes where the requirement is for the machine or process to follow a set sequence of operations. One example is a transfer line wherein each station performs a certain operation on a part, after which the processed part is transferred to the next station and replaced by an unprocessed part from the precious station. Another example is a process in which various dry bulk materials are weighed, combined, mixed and ultimately discharged.

Sequential control can be implemented in a number of ways, including electromechanical relays and various pneumatic, fluidic and solid-state devices. This chapter, however, focuses on digital computer systems that are primarily dedicated to the implementation of switching control. These special-purpose computers are called programmable logic control.

In light of the preceding discussion, it is clear that the role of the control unit is to continuously scan its program, setting outputs on or off depending on the switch conditions of the inputs. The controller must therefore be provided with a program that defines the desired switching sequences.

Notes

1. With each module having sixteen "points" of either input or output, this PLC has the ability to monitor and control dozens of devices. 由 With…所引导的短语形成逻辑主谓结构，作状语。本句可译为：PLC 中的每个模块都有 16 个"点"的输入或输出，具有监视和控制数十台装置的能力。

2. The relative simplicity of switching control makes it attractive for use in automatic machines and processes where the requirement is for the machine or process to follow a set sequence of operations. 在本句中，if 是代词，代替 switching control。本句可译为：在自动机床和加工过程中，相对简单的转换控制很有吸引力，因为这些机床和加工过程要遵循一套操作程序。

New Words

capability [ˌkeipəˈbiliti]　n. 性能
timing [ˈtaimiŋ]　n. 定时
drum [drʌm]　n. 鼓
sequencing [ˈsiːkwənsiŋ]　n. 测序
reliability [riˌlaiəˈbiliti]　n. 可靠性
module [ˈmɔdjuːl]　n. 模块
monitor [ˈmɔnitə]　v. 监视
cabinet [ˈkæbinit]　n. 柜
duplicate [ˈdjuːplikeit]　v. 复制
network [ˈnetwəːk]　n. 网络
special-purpose [ˈspeʃəlˈpəːpəs]　a. 专用的
image [ˈimidʒ]　n. 图像
treatment [ˈtriːtmənt]　n. 处理
characteristic [ˌkæriktəˈristik]　n. 特征
push-button [puʃˈbʌtn]　n. 按钮
solenoid [ˈsəulinɔid]　n. 螺线管
indicator [ˈindikeitə]　n. 显示器
energize [ˈenədʒaiz]　v. 激活
de-energize [diˈenədʒaiz]　v. 切断……为电源，使去能
statement [ˈsteitmənt]　n. 语句

declaration [ˌdekləˈreiʃən]　n. 陈述
structure [ˈstrʌktʃə]　v. 组成
transfer [trænsˈfəː]　v./n. 传送
bulk [bʌlk]　n. 大量
implement [ˈimplimənt]　v. 实现
pneumatic [nju(ː)ˈmætik]　a. 气体的，气动的
fluidic [fluːˈidik]　a. 流体的
scan [skæn]　v. 扫描

Phrases and Expressions

Programmable Logic Controller (PLC)　可编程逻辑控制器
time-delay relay　时滞继电器
drum sequencing　鼓测序
control cabinet　控制柜
take up　占用
pumping station　泵站
switching device　开关装置
limit switch　限位开关
on the basis of　基于
be required to　用于
be comprised of　由……组成
function block (FB)　功能模块
organization block (OB)　组织模块
program block (PB)　程序模块
sequence block (SB)　顺序模块
data block (DB)　数据模块
be available for　用于
user program　用户程序
go through　完成
give rise to　产生
be dedicated to　专门用于
in light of　鉴于
ladder diagram　梯形图

EXERCISE 1

The following is set of symbols denoting Working Safety. Choose the best symbol according to the information given below.

Unit 10 Introduction to Programmable Logic Controller

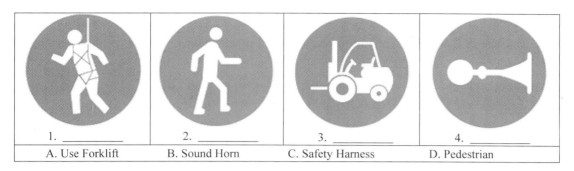

| 1. _____ | 2. _____ | 3. _____ | 4. _____ |
| A. Use Forklift | B. Sound Horn | C. Safety Harness | D. Pedestrian |

EXERCISE 2

Translate the following phrases into Chinese or English.

1. time-delay relay   _____
2. _____   开关装置
3. function block   _____
4. _____   程序模块
5. limit switch   _____
6. _____   模块
7. ladder diagram   _____
8. _____   用户程序

Passage B Connection of Programmable Logic Controllers

Signal connection and programming standards vary somewhat between different models of PLC, but they are similar enough to allow a "generic" introduction to PLC programming here. The following illustration (Figure 10-5) shows a simple PLC, as it might appear from a front view. Two screw terminals provide connection to 120 volts AC for powering the PLC's internal circuitry, labeled L1 and L2. Six screw terminals on the left-hand side provide connection to input devices, each terminal representing a different input "channel" with its own "X" label. The lower-left screw terminal is a "common"

Figure 10-5

connection, which is generally connected to L2 (neutral) of the 120 VAC power source.

Inside the PLC housing, connected between each input terminal and the common terminal, is an opto-isolator device (Light-Emitting Diode) that provides an electrically isolated "high" logic signal to the computer's circuitry (a photo-transistor interprets the LED's light) when there is 120 VAC power applied between the respective input terminal and the Common terminal. An indicating LED on the front panel of the PLC gives visual indication of an "energized" input (Figure 10-6).

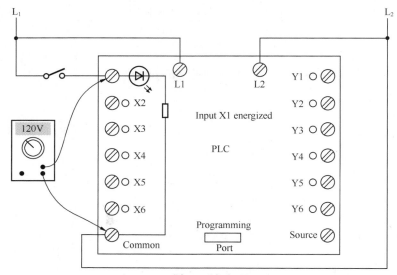

Figure 10-6

Output signals are generated by the PLC's computer circuitry activating a switching device (transistor, TRIAC, or even an electromechanical relay), connecting the "Source" terminal to any of the "Y" labeled output terminals. The "Source" terminal, correspondingly, is usually connected to the L1 side of the 120 VAC power source. As with each input, an indicating LED on the front panel of the PLC gives visual indication of an "energized" output (Figure 10-7).

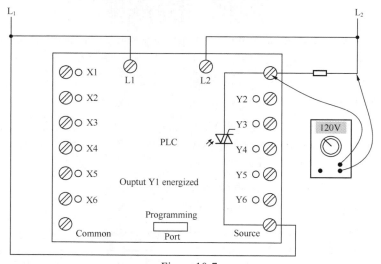

Figure 10-7

Unit 10 Introduction to Programmable Logic Controller

In this way, the PLC is able to interface with real-world devices such as switches and solenoids.

The actual logic of the control system is established inside the PLC by means of a computer program. This program dictates which output gets energized under which input conditions. Although the program itself appears to be a ladder logic diagram, with switch and relay symbols, there are no actual switch contacts or relay coils operating inside the PLC to create the logical relationships between input and output. These are imaginary contacts and coils, if you will. The program is entered and viewed via a personal computer connected to the PLC's programming port.

Consider the following circuit and PLC program (Figure 10-8).

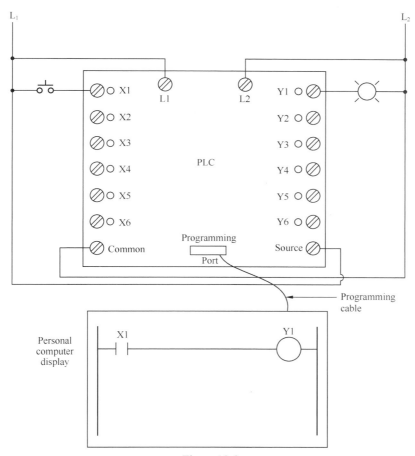

Figure 10-8

When the push-button switch is unactuated (unpressed), no power is sent to the X1 input of the PLC. Following the program, which shows a normally-open X1 contact in series with a Y1 coil, no "power" will be sent to the Y1 coil. Thus, the PLC's Y1 output remains de-energized, and the indicator lamp connected to it remains dark.

If the push-button switch is pressed, however, power will be sent to the PLC's X1 input. Any and all X1 contacts appearing in the program will assume the actuated (non-normal) state, as though they were relay contacts actuated by the energizing of a relay coil named "X1". In this case, energizing the X1 input will cause the normally-open X1 contact will "close", sending "power" to the Y1 coil. When the Y1 coil of the program "energizes", the real Y1 output will become energized, lighting up the lamp connected to it (Figure 10-9).

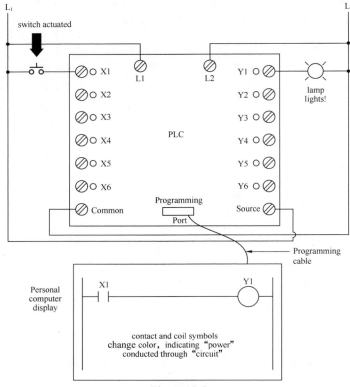

Figure 10-9

It must be understood that the X1 contact, Y1 coil, connecting wires, and "power" appearing in the personal computer's display are all virtual. They do not exist as real electrical components. They exist as commands in a computer program—a piece of software only—that just happens to resemble a real relay schematic diagram.

Furthermore, since each output in the PLC is nothing more than a bit in its memory as well, we can assign contacts in a PLC program "actuated" by an output (Y) status. Take for instance this next system, a motor start-stop control circuit (Figure 10-10).

The pushbutton switch connected to input X1 serves as the "Start" switch, while the switch connected to input X2 serves as the "Stop", Another contact in the program, named Y1, uses the output coil status as a seal-in contact, directly, so that the motor contactor will continue to be energized after the "Start" push-button switch is released. You can see the normally-closed contact X2 appear in a colored block, showing that it is in a closed ("electrically conducting") state.

If we were to press the "Start" button, input X1 would energize, thus "closing" the X1 contact in the program, sending "power" to the Y1 coil, energizing the Y1 output and applying 120 volt AC power to the real motor contactor coil. The parallel Y1 contact will also "close", thus latching the "circuit", in an energized state (Figure 10-11).

Unit 10 Introduction to Programmable Logic Controller

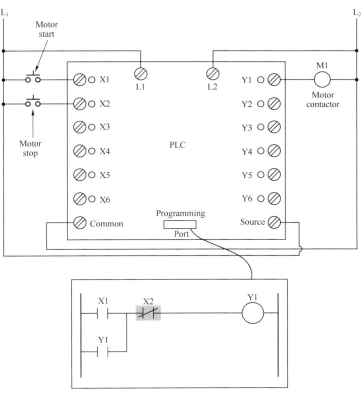

Figure 10-10

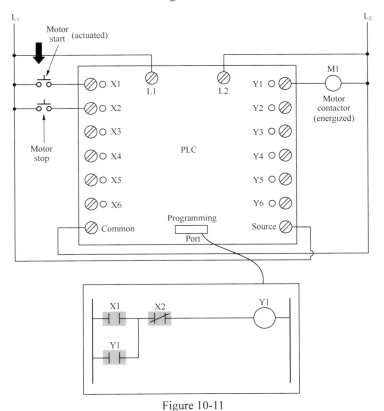

Figure 10-11

Now, if we release the "Start" push-button, the normally-open X1 contact will return to its "Open" state, but the motor will continue to run because the Y1 seal-in contact continues to provide "continuity" to "power" coil Y1, thus keeping the Y1 output energized.

To stop the motor, we must momentarily press the "Stop" push-button, which will energize the X2 input and "open" the normally-closed contact, breaking continuity to the Y1 coil.

When the "Stop" push-button is released, input X2 will de-energize, returning contact X2 to its normal, "closed" state. The motor, however, will not start again until the "Start" push-button is actuated, because the seal-in of Y1 has been lost.

New Words

model ['mɔdl]　n. 型号
circuitry ['sə:kitri]　n. 电路
label ['leibl]　v. 标有
channel ['tʃænl]　n. 通道
common ['kɔmən]　n. 公共端
opto-isolator [ˌɔptə'aisəˌleitə]　n. 光电隔离器
apply [ə'plai]　v. 施加
dictate [dik'teit]　v. 产生命令
cable ['keibl]　n. 电缆
normally-open ['nɔ:məli'əupən]　a. 常开的
non-normal ['nɔn'nɔ:məl]　a. 常闭的
virtual ['və:tjuəl]　a. 虚拟的
assign [ə'sain]　v. 分配
status ['steitəs]　n. 状态
contactor ['kɔntæktə]　n. 接触器
seal-in [si:lin]　a. 自锁的
release [ri'li:s]　v. 释放
normally-closed ['nɔ:məlikləuzd]　a. 常闭的
latch [lætʃ]　v. 锁定
continuity [ˌkɔnti'nju(:)iti]　n. 连续性
momentarily ['məuməntərili]　ad. 立刻

Phrases and Expressions

front view　正视图
TRIAC　双向可控硅
normally-open contact　常开触点
non-normal contact　常闭触点

be in series with 与……串联
schematic diagram   n. 框图
serve as 作为
seal-in contact 自锁触点
normally-closed contact 常闭触点

EXERCISE 3

Choose the best electronic accessory according to the information given.

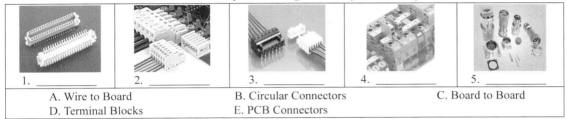

1. _____   2. _____   3. _____   4. _____   5. _____

A. Wire to Board B. Circular Connectors C. Board to Board
D. Terminal Blocks E. PCB Connectors

EXERCISE 4

Abbreviations are very useful in practical work. Read them and then translate them into corresponding Chinese terms.

1. SV	Safety Valve	_____	5. TR	Technical Regulation	_____
2. TB	Tee-bend	_____	6. TS	Tool Steel	_____
3. T.O.	Technical Order	_____	7. VHF	Very High Frequency	_____
4. TPI	Teeth Per Inch	_____	8. WB	Wheel Base	_____

Part II   Glance at Programmable Logic Control

The following is the structure of programmable logical controller.

PLC-System Overview

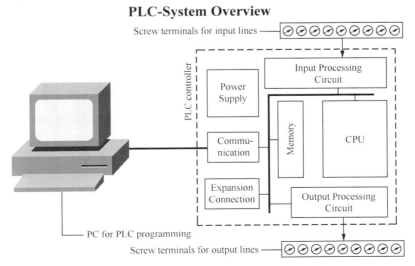

Explanations of the Programmable Logic Controller Terms

No.	Name	Explanation
1	PC for PLC programming	用于PLC编程的个人计算机
2	Screw terminals for input lines	输入螺丝接线端子排
3	Screw terminals for output lines	输出螺丝接线端子排
4	Power Supply	电源
5	Communication	通信
6	Expansion Connection	拓展连接
7	Input Processing Circuit	输入处理电路
8	Output Processing Circuit	输出处理电路
9	Memory	存储器
10	CPU	中央处理器

EXERCISE 5

The following is the typical control panel using PLC and local operator interface. You are required to choose the suitable term for each part.

Circuit Breakers	PLC	Digital Output Cards	Digital Input Cards
Analog Input Cards	Relay Switches	Power Supply	Nema 12 Enclosure
Operator Interface Terminal	Transient Surge Protectors		

1. _____   2. _____
3. _____   4. _____
5. _____   6. _____
7. _____   8. _____
9. _____   10. _____

Unit 10　Introduction to Programmable Logic Controller

| Part III　Simulated Writing |

Passage A　Match Your Skill

The following is a PLC, and you can learn the names of its parts.

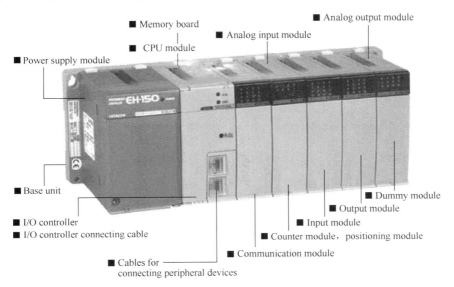

Explanations of the Programmable Logic Controller Terms

No.	Name	Explanation
1	Analog output module	模拟输出模块
2	Analog input module	模拟输入模块
3	Memory board	存储器板
4	CPU module	CPU 模块
5	Power supply module	电源模块
6	Base unit	底板
7	I/O controller	输入/输出控制器
8	I/O controller connecting cable	输入/输出控制器连接电缆
9	Cables for connecting peripheral devices	连接外设的电缆
10	Communication module	通信模块
11	Counter module, positioning module	计数、定位模块
12	Input module	输入模块
13	Output module	输出模块
14	Dummy module	空模块

EXERCISE 6

Match the words or phrases on the left with their meanings on the right.

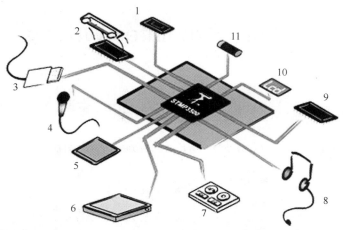

1. Host Processor (Optional)
2. FM Tuner
3. Hi-Speed USB
4. Microphone Voice Record
5. Flash Memory
6. Hard Drive
7. Buttons/Switches
8. Headphone
9. SDRAM
10. LED/LCD Screen
11. Rechargeable Battery

A. 按钮/开关
B. 硬盘驱动器
C. 可充电电池
D. 发光二级管/液晶显示器
E. 调频调谐器
F. 话筒录音
G. 同步动态随机存储器
H. 主处理器（可选）
I. 耳机
J. 闪存
K. 高速通用串行总线

Section B Have a Try

This section will help you to understand several forms of machining.

External Operations

Radial depth of cut—The depth of the tool along the radius of the work-piece as it makes a cut, as in a turning or boring operation. A large radial depth of cut will require a low feed rate, or else it will result in a high load on the tool and reduce the tool life. Therefore, a feature is often machined in several steps as the tool moves over at the radial depth of cut.

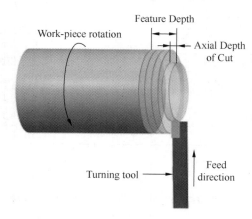

Unit 10 Introduction to Programmable Logic Controller

EXERCISE 7

This section is to test your ability to identify different operations.

| End milling (Slot) | End milling (Pocket) | End milling (Profile) |

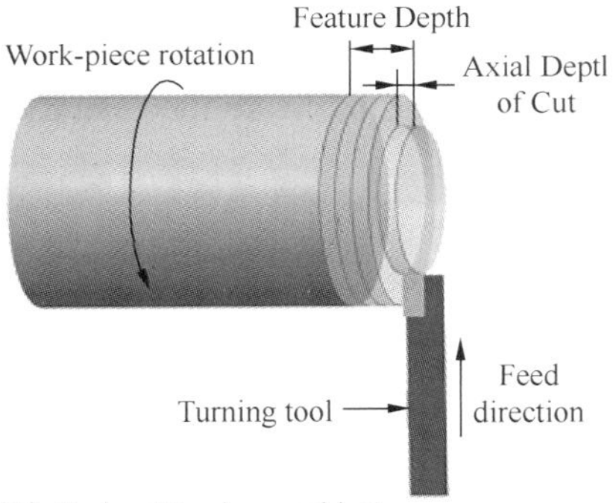

(1) Facing (Turning machine)

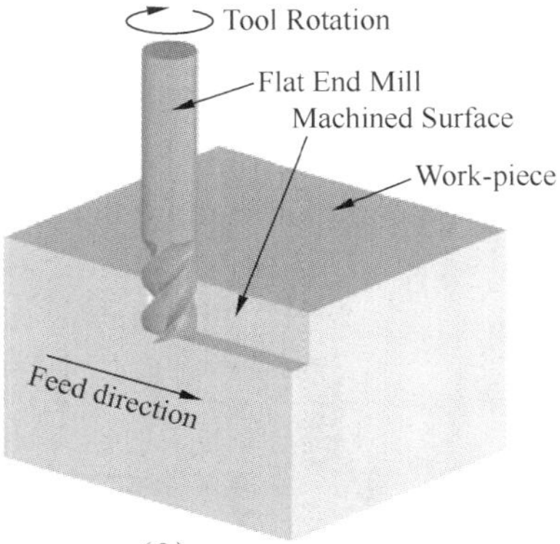

(2)

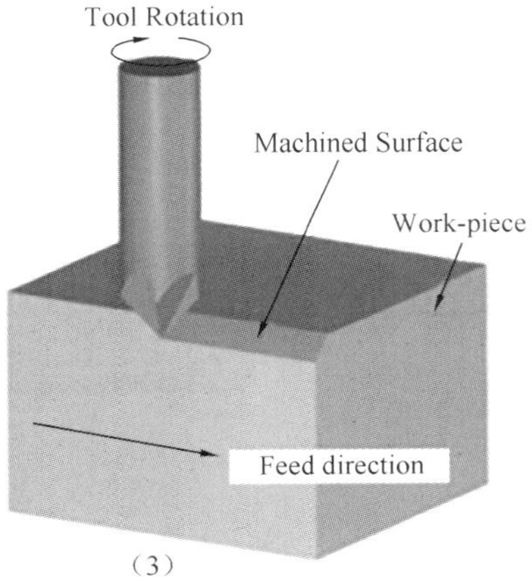

(3)

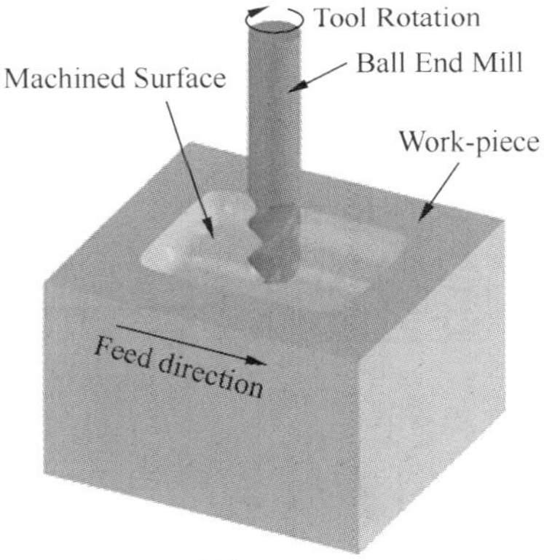

(4)

(1) Facing (Turning machine)
(2) _____
(3) _____
(4) _____

| Part IV   Broaden Your Horizon—
Practical Activity |

PLC Process Control System

A Programmable Logic Controller (PLC) is an industrial computer control system that continuously monitors the state of input devices and makes decisions based upon a custom program, to control the state of devices connected as outputs.

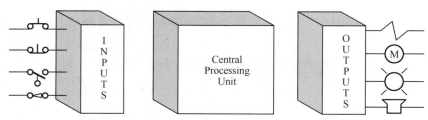

Almost any production line, machine function or process can be automated using a PLC. The speed and accuracy of the operation can be greatly enhanced using this type of control system. But the biggest benefit in using a PLC is the ability to change and replicate the operation or process while collecting and communicating vital information.

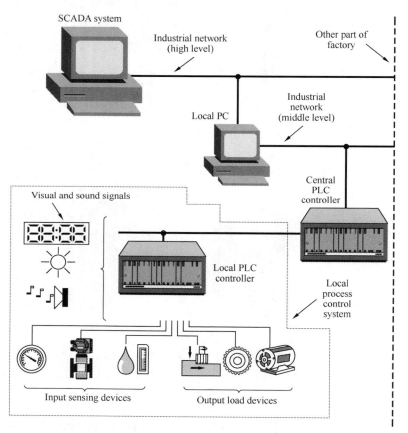

A process control system is made up of a group of electronic devices that provide stability, accuracy and eliminate harmful transition statuses in production processes. Operating systems can have different arrangements and implementations, from energy supply units to machines. As technology quickly progresses, many complex operational tasks have been solved by connecting programmable logic controllers and a central computer. Besides connections with devices (e.g., operating panels, motors, sensors, switches, valves, etc.), possibilities for communication among instruments are so great that they allow a high level of exploitation and process coordination. In addition, there is greater flexibility in realizing a process control system. Each component of a process control system plays an important role, regardless of its size. For example, without a sensor, the PLC wouldn't know what is going on during a process. In an automated system, a PLC controller is usually the central part of a process control system. With the execution of a program stored in program memory, PLC continuously monitors status of the system through signals from input devices. Based on the logic implemented in the program, PLC determines which actions need to be executed with output instruments. To run more complex processes it is possible to connect more PLC controllers to a central computer. A system could look like the one pictured below.

Glossary

absorb [əbˈsɔ:b]　v. 吸收　　8(B)
accessory [ækˈsesəri]　n. 附件　　1(A)
academic [ˌækəˈdemik]　a. 学术的　　6(A)
accelerate [ækˈseləreit]　v. 加速　　6(B)
access [ˈækses]　n. 存储，进入　　8(A)
accuracy [ˈækjurəsi]　n. 精度　　5(A)
accurate [ˈækjurit]　a. 精确的　　1(A)
acoustic [əˈku:stik]　a. 声的　　2(A)
activate [ˈæktiveit]　v. 激活，刺激　　5(A)
activation [ˌæktiˈveiʃən]　n. 启动，激活　　5(A)
actuate [ˈæktjueit]　v. 开动（机器），使运转　　4(A)
actuator [ˈæktjueitə]　n. 执行元件，执行装置，执行机构　　4(A)
adjustment [əˈdʒʌstmənt]　n. 调整　　1(B)
advance [ədˈvɑ:ns]　v. 向前运动　　3(B)
aerospace [ˈɛərəuspeis]　n. 航空航天　　2(A)
allowance [əˈlauəns]　n. 余量　　2(B)
alloy [ˈælɔi]　n. 合金　　1(A)
alter [ˈɔ:ltə]　v. 改变　　3(B)
alternating [ɔ:lˈtə:nitiŋ]　a. 交替的　　7(A)
alternatively [ɔ:lˈtɜ:nətivli]　ad. 作为选择　　5(B)
aluminum [əˈlju:minəm]　n. 铝　　1(A)
ampere [ˈæmpeə(r)]　n. 安培　　7(B)
amplifier [ˈæmpliˌfaiə]　n. 放大器　　7(A)
analog-to-digital [ˈænəlɔgtuˈdidʒitl]　a. 模数的　　8(A)
analogy [əˈnælədʒi]　n. 模拟　　4(A)
angle [ˈæŋgl]　n. 角度　　3(B)
anneal [əˈni:l]　v. 退火　　4(B)
anvil [ˈænvil]　n. 固定爪　　2(A)
appliance [əˈplaiəns]　n. 家用电器　　2(A)
apply [əˈplai]　v. 施加　　10(B)
approach [əˈprəutʃ]　n. 接近　　5(B)
arbor [ˈɑ:bə]　n. 刀轴　　1(A)
arc [ɑ:k]　n. 弧　　6(B)

architectural [ˌɑːkiˈtektʃərəl]   a. 建筑的   6(B)
architecture [ˈɑːkitektʃə]   n. 结构   8(A)
armature [ˈɑːmətjuə]   n. 电枢   9(A)
arrangement [əˈreindʒmənt]   n. 排列   4(A)
arrow [ˈærəu]   n. 箭头   7(B)
articulated [ɑːˈtikjulitid]   a. 铰链的   6(B)
assemble [əˈsembl]   v. 组装   9(B)
assign [əˈsain]   v. 分配   10(B)
attach [əˈtætʃ]   v. 固定   1(A)
attachment [əˈtætʃmənt]   n. 夹具   1(B)
attain [əˈtein]   v. 获得，达到   1(B)
attraction [əˈtrækʃən]   n. 吸引   9(A)
automate [ˈɔːtəmeit]   v. 使自动化   5(A)
available [əˈveiləbl]   a. 可用的，有空的   3(A)
axle [ˈæksl]   n. 车轴，轮轴   1(A)
background [ˈbækɡraund]   n. 背景   6(A)
band [bænd]   n. 带状   1(B)
base [beis]   n. 基极   8(B)
batch [bætʃ]   n. 一组   5(B)
bearing [ˈbɛəriŋ]   n. 轴承   1(B)
bell [bel]   n. 罩   5(A)
belt [belt]   n. 传送带   4(A)
bench [bentʃ]   n. 工作台   1(B)
bend [bend]   n. 弯曲成形件   2(B)
bevel [ˈbevəl]   n. 倾斜   2(B)
bias [ˈbaiəs]   v. 偏置   7(A)
bistable [baiˈsteibl]   a. 双稳态的   7(A)
blade [bleid]   n. 刀片   2(A)
blank [blæŋk]   n. 毛坯   2(B)
blow [bləu]   n. 锤打   4(B)
blueprint [ˈbluːˌprint]   n. 蓝图，设计图   3(A)
bolt [bəult]   n. 螺栓   1(A)
bore [bɔː]   v. 钻孔   3(A)
bracket [ˈbrækit]   n. 支架   2(B)
branch [brɑːntʃ]   n. 支路   7(B)
brass [brɑːs]   n. 黄铜   1(A)
broach [brəutʃ]   v. 拉削   3(A)
browser [ˈbrauzə(r)]   n. 浏览器   5(B)
brush [brʌʃ]   n. 电刷   9(A)

| 183 |

单词	页码
bulk [bʌlk]　n.　大量	10(A)
bus [bʌs]　n.　总线	8(A)
bush [buʃ]　n.　衬套	9(B)
bushing [ˈbuʃiŋ]　n.　隔套, 套管	4(A)
buzzing [ˈbʌziŋ]　a.　嗡嗡的	8(B)
buzzle [bʌzl]　n.　蜂鸣器	8(A)
cabinet [ˈkæbinit]　n.　柜	10(A)
cable [ˈkeibl]　n.　电缆	10(B)
caliper [kælˈpə]　n.　卡尺	1(A)
capability [ˌkeipəˈbiliti]　n.　性能	10(A)
capacitor [kəˈpæsitə]　n.　电容器	7(A)
carbon [ˈkɑːbən]　n.　碳	1(A)
carriage [ˈkæridʒ]　n.　滑鞍, 滑座	1(A)
cast [kɑːst]　n./v.　铸造	1(A)
cavity [ˈkæviti]　n.　腔, 型腔	2(B)
center [ˈsentə]　n.　顶尖	1(A)
centerline [ˈsentəlain]　n.　中线	1(B)
chain [tʃein]　n.　链条	2(A)
chamfer [ˈtʃæmfə]　v.　倒角, 倒槽	5(A)
changeover [ˈtʃeindʒˈəuvə]　n.　转换	7(A)
channel [ˈtʃænl]　n.　通道	10(B)
characteristics [ˌkæriktəˈristik]　n.　特性	2(A)
characterize [ˈkæriktəraiz]　v.　以……为特性	2(B)
charge [tʃɑːdʒ]　n.　电荷　v.　充电	7(A)
chassis [ˈʃæsi]　n.　底盘	2(B)
chip [tʃip]　n.　铁屑, 碎屑, 芯片	1(A)
chisel [ˈtʃizl]　n.　凿子	2(A)
chromium [ˈkrəumjəm]　n.　铬	2(A)
chuck [tʃʌk]　n.　夹具, 卡盘	1(A)
circuit [ˈsəːkit]　n.　回路, 电路	4(A)
circuitry [ˈsəːkitri]　n.　电路	10(B)
circular [ˈsəːkjulə]　a.　圆形的, 循环的	1(B)
clamp [klæmp]　v.　夹紧	1(A)
classify [ˈklæsifai]　v.　分类	3(B)
clearance [ˈkliərəns]　n.　间隙	5(B)
clockwise [ˈklɔkwaiz]　ad.　顺时针	9(A)
coat [kəut]　v.　涂料	4(B)
coating [ˈkəutiŋ]　n.　涂料	9(B)
cogging [ˈkɔgiŋ]　n.　钝齿啮合	4(B)

coil [kɔil] n. 线圈	7(A)
collector [kə'lektə] n. 集电极	8(B)
collet ['kɔlit] n. 棘爪	3(B)
command [kə'mɑːnd] n. 指令	3(B)
common ['kɔmən] n. 公共端	10(B)
commonality [,kɔmə'næliti] n. 共性，通用性	5(A)
commutator ['kɔmjuteitə] n. 换向器	9(A)
compass ['kʌmpəs] n. 圆规	6(B)
complex ['kɔmpleks] a. 复杂的	7(B)
complexity [kəm'pleksiti] n. 复杂(性)，复杂的事物	5(A)
component [kəm'pəunənt] n. 部件	7(A)
component [kəm'pəunənt] n. 元件，组件 v. 组成的	4(A)
composition [kɔmpə'ziʃən] n. 成分，组成	2(A)
compound ['kɔmpaund] n. 复合刀架	3(B)
compress [kəm'pres] v. 压缩	1(A)
compression [kəm'preʃ(ə)n] n. 压缩	2(B)
compressive [kəm'presiv] a. 压缩的	4(B)
compute [kəm'pjuːt] v. 计算	8(A)
conduct ['kɔndʌkt] v. 传导，动作	2(A)
conductance [kən'dʌktəns] n. 电导率	7(B)
conductivity [,kɔndʌk'tiviti] n. 传导性	2(A)
conductor [kən'dʌktə] n. 导体	7(A)
constant ['kɔnstənt] a. 恒定的	2(B)
consumption [kən'sʌmpʃən] n. 消耗	7(A)
contact ['kɔntækt] n. 触点	7(A)
contactor ['kɔntæktə] n. 接触器	10(B)
containment [kən'teinmənt] n. 控制	5(B)
context ['kɔntekst] n. 背景	6(A)
continuity [,kɔnti'nju(ː)iti] n. 连续性	10(B)
contoured ['kɔntuəd] a. 波状外形的	2(B)
convention [kən'venʃən] n. 常规	6(A)
conventional [kən'venʃənl] a. 常规的	3(A)
converge [kən'vəːdʒ] v. 汇聚	7(B)
convert [kən'vəːt] v. 转换	9(A)
converter [kən'vəːtə(r)] n. 转换器	8(A)
conveyer [kən'veiə] n. 输送机	4(B)
cookware ['kukweər] n. 炊具	2(A)
coolant ['kuːlənt] n. 冷却剂	5(A)

单词	课次
coordinate [kəuˈɔ:dinit] v. 协调	5(A)
core [kɔ:] n. 铁芯，外存	5(A)
cored [kɔ:d] a. 型芯的	4(B)
corrosion [kəˈrəuʒən] n. 腐蚀	2(A)
corrosive [kəˈrəusiv] a. 腐蚀的	2(A)
counterbalance [ˌkauntəˈbæləns] n. 平衡	4(A)
counterclockwise [ˌkauntəˈklɔkwaiz] ad. 逆时针	9(A)
coupling [ˈkʌpliŋ] n. 联轴器	4(A)
crankshaft [ˈkræŋkʃɑ:ft] n. 曲轴	2(A)
crystal [ˈkristl] n. 晶体	8(A)
current-carrying [ˈkʌrəntˈkæriiŋ] a. 载流的	9(B)
curve [kə:v] n. 曲线	6(B)
curved [kə:vd] a. 弯曲的	2(B)
customizable [kʌstəmaiz] a. 可定制化的	5(B)
cutlery [ˈkʌtləri] n. 餐具	2(A)
cutout [ˈkʌtaut] v. 切除，切去	6(A)
cutter [ˈkʌtə] n. 刀具	3(A)
cylinder [ˈsilində] n. 汽缸，圆柱	1(B)
cylindrical [siˈlindrik(ə)l] a. 圆柱的	3(A)
datasheet [ˈdeitəʃi:t] n. 数据手册	8(B)
decent-quality [ˈdi:sntˈkwɔliti] a. 优质的	1(A)
declaration [ˌdekləˈreiʃən] n. 陈述	10(A)
dedicated [ˈdedikeitid] a. 专用的	1(B)
de-energize [diˈenədʒaiz] v. 切断……电源	10(A)
definition [ˌdefiˈniʃən] n. 精确度	5(B)
deform [di:ˈfɔ:m] v. 变形	4(B)
deformation [ˌdi:fɔ:ˈmeiʃən] n. 变形	2(B)
density [ˈdensiti] n. 密度	1(B)
demonstrate [ˈdemənstreit] v. 演示	8(B)
department [diˈpɑ:tmənt] n. 部	8(A)
derivative [diˈrivətiv] n. 衍生品	8(A)
designate [ˈdezigneit] v. 指定	7(B)
diagram [ˈdaiəgræm] n. 图	7(B)
dial [ˈdaiəl] n. 刻度表	1(A)
diameter [daiˈæmitə] n. 直径	1(A)
dictate [dikˈteit] v. 产生命令	10(B)
die [dai] n. 模具	2(B)
die-casting [daiˈkɑ:stiŋ] n. 压铸件	2(A)
dielectric [ˌdaiiˈlektrik] n. 电介质	7(A)

digital ['didʒitl]　a. 数字的　　　　　　　　　　　　　　　　　　1(A)
dimension [di'menʃən]　n. 尺寸　　　　　　　　　　　　　　　　1(A)
dimensioning [di'menʃəniŋ]　n. 标注尺寸　　　　　　　　　　　6(A)
ding [diŋ]　v. 发出声响　　　　　　　　　　　　　　　　　　　1(B)
diode ['daiəud]　n. 二极管　　　　　　　　　　　　　　　　　　7(A)
directory [di'rektəri]　n. 目录　　　　　　　　　　　　　　　　5(B)
disassembly [,disə'sembli]　n. 拆卸　　　　　　　　　　　　　6(A)
disc [disk]　n. 圆盘形　　　　　　　　　　　　　　　　　　　　2(B)
discharge [dis'tʃɑːdʒ]　n. 放电　　　　　　　　　　　　　　　3(A)
discipline ['disiplin]　n. 科目　　　　　　　　　　　　　　　　6(A)
discrete [dis'kriːt]　a. 离散的　　　　　　　　　　　　　　　　7(A)
displacement [dis'pleismənt]　n. 排量　　　　　　　　　　　　4(A)
distribute [dis'tribju(ː)t]　v. 分布　　　　　　　　　　　　　　9(A)
drafter ['drɑːftə]　n. 制图员　　　　　　　　　　　　　　　　　6(A)
drafting ['drɑːftiŋ]　n. 制图　　　　　　　　　　　　　　　　　6(A)
draftsperson [drɑːfts'pəːsn]　n. 绘图员　　　　　　　　　　　6(A)
drain [drein]　n. 泄油　　　　　　　　　　　　　　　　　　　　4(A)
drawback ['drɔːˌbæk]　n. 弊端　　　　　　　　　　　　　　　　1(B)
drill [dril]　n. 钻头
　　　　　　v. 钻　　　　　　　　　　　　　　　　　　　　　　1(A)
drum [drʌm]　n. 鼓　　　　　　　　　　　　　　　　　　　　　10(A)
ductile ['dʌktail]　a. 易延展的，有韧性的　　　　　　　　　　 2(A)
ductility [dʌk'tiliti]　n. 延展性　　　　　　　　　　　　　　　4(B)
duplicate ['djuːplikeit]　v. 复制　　　　　　　　　　　　　　　10(A)
edge [edʒ]　n. 刀口　　　　　　　　　　　　　　　　　　　　　3(A)
elastic [i'læstik]　a. 有弹性的　　　　　　　　　　　　　　　　2(B)
electro-chemical [i,lektrəu'kemikəl]　a. 电化学的　　　　　　3(A)
electrode [i'lektrəud]　n. 电极板，电极棒　　　　　　　　　　1(A)
electroluminescence [i'lektrəuˌljuːmi'nesəns]　n. 电致发光　　7(A)
electromagnetic [ilektrəu'mægnitik]　a. 电磁的　　　　　　　9(A)
electromechanical　a. [i,lektrəumiˈkænikəl]　机电的　　　　　9(A)
electromotive [ilektrəu'məutiv]　a. 电动的　　　　　　　　　 7(B)
electron [i'lektrɔn]　n. 电子　　　　　　　　　　　　　　　　　7(A)
electronic [ilek'trɔnik]　a. 电子的　　　　　　　　　　　　　　7(A)
electronics [ilek'trɔniks]　n. 电子，电子学　　　　　　　　　　7(A)
element ['elimənt]　n. 元件　　　　　　　　　　　　　　　　　 7(A)
emanate ['eməneit]　v. 发源，发生　　　　　　　　　　　　　 6(A)
embed [im'bed]　v. 嵌入　　　　　　　　　　　　　　　　　　 8(A)
emit [i'mit]　v. 发射　　　　　　　　　　　　　　　　　　　　 7(A)

| 187 |

emitter [iˈmitə] n. 发射极	8(B)
enclosure [inˈkləuʒə] n. 外壳	2(B)
encoder [inˈkəudə] n. 编码器	5(A)
encryption [inˈkripʃən] n. 加密	8(B)
endbell [endbel] n. 端承口	9(B)
energize [ˈenədʒaiz] v. 激活	10(A)
engage [inˈgeidʒ] n. 啮合	5(B)
enlarge [inˈlɑːdʒ] v. 扩大	3(A)
equation [iˈkweiʃən] n. 方程	9(A)
erosion [iˈrəuʒən] n. 腐蚀	3(A)
exceed [ikˈsiːd] v. 超过	8(A)
excess [ikˈses, ˈekses] a. 多余的	5(B)
exchanger [iksˈtʃeindʒə] n. 交换器	2(A)
execute [ˈeksikjuːt] v. 执行	5(A)
explicit [iksˈplisit] a. 明确的	7(B)
explore [iksˈplɔː] v. 探索	5(B)
explorer [iksˈplɔːrə] n. 资源管理器	5(B)
extend [iksˈtend] v. 拉出	4(A)
external [eksˈtəːnl] a. 外部的	8(A)
extremity [iksˈtremiti] n. 末端	3(A)
fabrication [ˌfæbriˈkeiʃən] n. 制造	2(A)
faceplate [ˈfeispleit] n. （车床）花盘	1(A)
factor [ˈfæktə] n. 系数	2(B)
fahrenheit [ˈfærənhait] n. 华氏度	4(B)
feed [fiːd] v. 供料，提供	3(B)
feedrate [fiːdəreit] n. 进给速度	5(A)
ferrous [ˈferəs] a. 含铁的	2(A)
field [fiːld] n. 场	9(B)
file [fail] n. 锉刀	2(A)
filter [ˈfiltə] n. 过滤器 v. 过滤	4(A)
finish [ˈfiniʃ] n. 抛光度	1(B)
fitting [ˈfitiŋ] n. 配件	2(A)
fix [fiks] v. 固定	9(B)
fixture [ˈfikstʃə] n. 夹具	3(B)
flange [flændʒ] n. 凸缘，法兰	2(B)
flash [flæʃ] n. 飞边，溢料	4(B)
flip-latch [flipˈlætʃ] n. 锁存器	8(A)
fluid [ˈfluː(ː)id] n. 流体	4(A)

fluidic [flu:`idik]   a. 流体的   10(A)
flux [flʌks]   n. 通量   9(A)
fold [fəuld]   v. 折   1(A)
foolproof ['fu:lpru:f]   a. 不会出错的   1(A)
foreground ['fɔ:graund]   n. 前景   6(A)
forge [fɔ:dʒ]   v. 锻造   4(A)
forging ['fɔ:dʒiŋ]   n. 锻件，锻造法   2(A)
form [fɔ:m]   v. 成形   2(A)
formability [fɔ:mə'biliti]   n. 成形性   2(A)
format ['fɔ:mæt]   n. 格式   5(A)
foundation [faun'deiʃən]   n. 基座   4(B)
fracture ['fræktʃə]   v. 断裂   4(B)
frame [freim]   n. 框   2(A)
freehand ['fri:hænd]   a. 凭手画的   6(B)
frequency ['fri:kwənsi]   n. 频率   8(B)
friction ['frikʃən]   n. 摩擦   1(B)
fundamental [,fʌndə'mentl]   a. 基本的   7(B)
gage [geidʒ]   n. 量规   1(A)
galvanize ['gælvənaiz]   v. 镀锌   2(A)
gate [geit]   n. 门极   7(A)
general-purpose ['dʒenərəl'pə:pəs]   a. 通用的   8(A)
generate ['tekstʃe]   v. 生成   3(A)
generator ['dʒenəreitə]   n. 发电机，发生器   7(B)
geometry [dʒi'ɔmitri]   n. 几何形状   3(A)
graphic ['græfik]   a. 图的，图形的   6(A)
grease [gri:s]   n. 润滑脂   1(A)
grind [graind]   v. 磨削   1(A)
grip [grip]   v. 夹住   2(B)
groove [gru:v]   n. 凹槽，环槽   4(A)
ground [graund]   n. 接地   8(B)
hammer ['hæmə]   n. 锤子   1(B)
hardness ['hɑ:dnis]   n. 硬度   2(A)
hardware ['hɑ:dweə]   n. 五金   2(A)
heal [hi:l]   v. 愈合   2(A)
heater ['hi:tə]   n. 加热机   4(B)
heat-treated ['hi:ttri:tid]   a. 热处理的   2(A)
hexagonal [hek'sægənəl]   a. 六角形的   1(B)
holder ['həuldə]   n. 刀架   1(B)
holder ['həuldə]   n. 固定装置   3(B)

hopper ['hɔpə]　　n. 装料斗　　　　　　　　　　　　　　　　4(B)
horizontal [ˌhɔri'zɔntl]　　a. 水平式的　　　　　　　　　　　3(B)
horizontally [ˌhɔri'zɔntli]　　ad. 沿水平方向　　　　　　　　2(B)
housing ['hauziŋ]　　n. 外壳，外罩　　　　　　　　　　　　　4(A)
hydraulic [hai'drɔ:lik]　　adj. 液压的，水压的　　　　　　　4(A)
hydraulics ['hai'drɔ:liks]　　n. 液压技术，水力学　　　　　　4(A)
hydrostatic [ˌhaidrəu'stætik]　　a. 静水力学的　　　　　　　4(A)
illustrate ['iləstreit]　　v. 图解，　图示　　　　　　　　　　9(A)
illustration [ˌiləs'treiʃən]　　n. 图解　　　　　　　　　　　6(A)
image ['imidʒ]　　n. 图像　　　　　　　　　　　　　　　　　10(A)
implement ['implimənt]　　v. 实现　　　　　　　　　　　　　10(A)
incidentally [ɪnsɪ'dentəlɪ]　　ad. 顺便说一句　　　　　　　　1(A)
increment ['inkrimənt]　　n. 增量　　　　　　　　　　　　　5(A)
independent [indi'pendənt]　　a. 独立的　　　　　　　　　　7(B)
indicator ['indikeitə]　　n. 显示器　　　　　　　　　　　　10(A)
in-house [in'haus]　　a. 内部的　　　　　　　　　　　　　　8(B)
initial [i'niʃəl]　　a. 最初的　　　　　　　　　　　　　　　5(B)
inlet ['inlet]　　n. 入口　　　　　　　　　　　　　　　　　4(A)
installation [ˌinstə'leiʃne]　　n. 安装　　　　　　　　　　　5(B)
instruction [in'strʌkʃən]　　n. 指令　　　　　　　　　　　　5(A)
insulator ['insjuleitə]　　n. 绝缘体　　　　　　　　　　　　7(A)
insulating ['insjuleitlŋ]　　v. 绝缘的　　　　　　　　　　　9(B)
integrate ['intigreit]　　v. 集成　　　　　　　　　　　　　8(A)
intensity [in'tensiti]　　n. 强度　　　　　　　　　　　　　7(B)
interact [ˌintər'ækt]　　v. 相互作用　　　　　　　　　　　　9(A)
interconnection [ˌintəkə'nekʃən]　　n. 相互连接　　　　　　8(A)
interface ['intə(:)ˌfeis]　　n. 接口，界面
　　　　　　　　　　　　v. 连接　　　　　　　　　　　　　　5(B)
interference [ˌintə'fiərəns]　　n. 干涉　　　　　　　　　　9(A)
internal [in'tə:nl]　　a. 内部的　　　　　　　　　　　　　　8(A)
interpret [in'tə:prit]　　v. 解译，图示，中断　　　　　　　　5(A)
interrupt　(INIT) [ˌintə'rʌpt]　　n. 中断　　　　　　　　　　8(B)
intersect [ˌintə'sekt]　　v. 相交于　　　　　　　　　　　　4(A)
jaw [dʒɔ:]　　n. 夹具　　　　　　　　　　　　　　　　　　1(B)
jet [dʒet]　　n. 喷射　　　　　　　　　　　　　　　　　　3(A)
keen [ki:n]　　a. 锋利的　　　　　　　　　　　　　　　　　2(A)
kitchenware ['kitʃinweə(r)]　　n. 厨具　　　　　　　　　　2(A)
knead [ni:d]　　v. 捏制　　　　　　　　　　　　　　　　　4(B)
label ['leibl]　　v. 标有　　　　　　　　　　　　　　　　　10(B)

lamination [ˌlæmiˈneiʃən] n. 叠片，铁芯片   5(A)

laser [ˈleizə] n. 激光   3(A)

latch [lætʃ] v. 锁定   10(B)

lathe [leið] n. 车床   1(A)

layer [ˈleiə] n. 层   2(A)

layout [ˈlei,aut] n. 布局   6(A)

lead [li:d] n. 导线，铅，引线   2(A)

leave [li:v] n. 金属片   1(A)

lever [ˈli:və, ˈlevə] n. 杠杆，控制杆，电平
                    v. 撬起   4(A)

limber [ˈlimbə] a. 柔性的，易弯曲的   1(A)

linear [ˈliniə] a. 线性的   5(A)

lithium [ˈliθiəm] n. 锂   1(B)

live [ləiv] a. 活动的，带电的   3(B)

load [ləud] n. 负载   7(B)

localized [ˈləukəlaizd] a. 定位的   4(B)

location [ləuˈkeiʃən] n. 位置   6(A)

lock-up [lɔkʌp] n. 锁紧   9(B)

logical [ˈlɔdʒikəl] a. 逻辑的   5(A)

lubricate [ˈlu:brikeit] v. 润滑   1(B)

lubrication [ˌlu:briˈkeiʃən] n. 润滑   5(A)

machinable [məˈʃi:nəbl] a. 可加工的   2(A)

machinery [məˈʃi:nəri] n. （总称）机器，机械   4(A)

machining [məˈʃi:niŋ] n. 加工   3(A)

machinist [məˈʃi:nist] n. 机械师   1(A)

magazine [ˌmæɡəˈzi:n] n. 机台   4(B)

magnesium [mæɡˈni:zjəm] n. 镁   2(A)

magnet [ˈmæɡnit] n. 磁铁   9(A)

magnetic [mæɡˈnetik] a. 有磁性的   2(A)

magnetize [ˈmæɡnitaiz] v. 磁化   9(A)

magnify [ˈmæɡnifai] v. 放大   7(A)

major [ˈmeidʒə] a. 大的   1(A)

malleable [ˈmæliəbl] a. 可塑的   2(A)

mandrel [ˈmændril] n. 心轴   2(A)

manganese [ˌmæŋɡəˈni:z] n. 锰   2(A)

manipulate [məˈnipjuleit] v. 操纵   5(A)

manual [ˈmænjuəl] a. 手动的   3(B)

manually [ˈmænjuəli] ad. 人工地   6(B)

manufacture [ˌmænjuˈfæktʃə] n. 制造，制造业

v. 制造	3(A)
margin ['mɑːdʒin] n. 页边距	6(B)
mask [mɑːsk] n. 口罩	1(A)
mate [meit] v. 匹配	1(B)
measurement ['meʒəmənt] n. 测量	1(A)
mechanically [mi'kænikəli] ad. 机械地	3(A)
mechanics [mi'kæniks] n. 技巧	6(B)
medium ['miːdiəm] n. 介质	4(A)
member ['membə] n. 构件	4(B)
memory ['meməri] n. 内存	8(A)
metallic [mi'tælik] a. 金属的	7(A)
metallurgical [ˌmetə'lɜːdʒikəl] a. 冶金的	4(B)
metric ['metrɪk] n. 米制，公制的	1(A)
microcontroller [kən'trəulə] n. 微控制器	8(A)
mill [mil] n. 铣床	
v. 铣	1(A)
minimize ['minimaiz] v. 减到最少	5(B)
minor ['mainə] a. 小的	1(A)
minus ['mainəs] a. 负的	9(A)
miscellaneous [misi'leinjəs] a. 其他的	3(A)
mobile ['məubail] a. 移动的	7(A)
model ['mɔdl] n. 型号	10(B)
modulation [ˌmɔdju'leiʃən] n. 调制	8(B)
module ['mɔdjuːl] n. 模块	10(A)
moisture ['mɔistʃə] n. 水分	1(B)
mold [məuld] n. 模子	2(B)
molybdenum [mə'libdinəm] n. 钼	2(A)
momentarily ['məuməntərili] ad. 立刻	10(B)
monitor ['mɔnitə] v. 监视	10(A)
motion ['məuʃən] n. 运动	9(A)
motor-housing ['məutə'hauziŋ] n. 电机壳	9(B)
mount [maunt] v. 安装	1(A)
mouth [mauθ] n. 浇口	4(B)
multiple ['mʌltipl] a. 多个的	3(A)
multiplication [ˌmʌltipli'keiʃən] n. 增加，倍增	4(A)
nail [neil] n. 钉子	2(A)
namely ['neimli] ad. 即	7(B)
nature ['neitʃə] n. 特性	9(A)
negate [ni'geit] v. 消除	4(B)

词条	位置
negative ['negətiv]　n. 负极	7(A)
network ['netwə:k]　n. 网络	10(A)
networking ['netwə:kiŋ]　n. 网络化	8(A)
neutral ['nju:trəl]　a. 中间的，中性线	2(B)
nickel ['nikl]　n. 镍	2(A)
node [nəud]　n. 节点	7(B)
no-load ['nəuləud]　n. 空载	9(B)
nomenclature [nəu'menklətʃə]　n. 术语	6(A)
nominal ['nɔminl]　a. 公称的，基本的	1(A)
noncombustible ['nɔnkəm'bʌstəbl]　a. 不燃的	2(A)
non-normal ['nɔn'nɔ:məl]　a. 常闭的	10(B)
non-uniform ['nɔn'ju:nifɔ:m]　a. 无统一形状的	2(B)
non-volatile ['nɔn'vɔlətail]　a. 永久性的	8(A)
normally-closed ['nɔ:məli kləuzd]　a. 常闭的	10(B)
normally-open ['nɔ:məli'əupən]　a. 常开的	10(B)
notation [nəu'teiʃən]　n. 记号，标记	6(A)
nozzle ['nɔzl]　n. 喷管	4(B)
numerical [nju(:)'merikəl]　a. 数字的	3(B)
odd-shaped [ɔdʃeɪpt]　a. 奇特形状的，不规则的	1(A)
offset ['ɔ:fset]　v. 偏置	1(A)
ohm [əum]　n. 欧姆	7(A)
opaque [əu'peik]　a. 不透明的	6(A)
operator ['ɔpəreitə]　n. 操作者	3(B)
option ['ɔpʃən]　n. 选项	5(B)
opto-coupler [,ɔptəu'kʌplə]　n. 光电耦合器	8(A)
opto-isolator [,ɔptə'aisə,leitə]　n. 光电隔离器	10(B)
orient ['ɔ:riənt]　v. 确定方向	2(B)
ornament ['ɔ:nəmənt]　n. 饰品	2(A)
oscillator ['ɔsileitə]　n. 示波器	7(A)
oscillator ['ɔsileitə]　n. 振荡器	8(A)
outlet ['autlet]　n. 插座	9(B)
over-bend ['əuvəbend]　v. 过度弯曲	2(B)
overly ['əuvəli]　adv. 过度地，极度地	5(A)
oxide ['ɔksaid]　n. 氧化物	2(A)
package ['pækidʒ]　v. 封装	
n. 封装	7(A)
pad [pæd]　n. 垫片	7(A)
panel ['pænl]　n. 板，面板	2(B)
parallel ['pærəlel]　a. 并联的，并行的	4(A)

parameter [pə'ræmitə] n. 参数	2(B)
penetration [peni'treiʃən] n. 划入	3(A)
peripheral [pə'rifərəl] a. 外围的	8(A)
n. 外设	
perpendicular [,pə:pən'dikjulə] a. 垂直的	3(A)
physical ['fizikəl] a. 实物的	6(A)
pin [pin] n. 管脚	8(A)
placement ['pleismənt] n. 位置	5(A)
plane [plein] v. 刨面	3(A)
platform ['plætfɔ:m] n. 平台	3(B)
plating ['pleitiŋ] n. 镀饰，电镀	2(A)
plug [plʌg] n. 塞子	1(A)
plus [plʌs] a. 正的	9(A)
pneumatic [nju(:)'mætik] a. 气体的，气动的	10(A)
pneumatics [nju:'mætiks] n. 气压工程，气体力学	4(A)
polarity [pəu'læriti] n. 极性	9(A)
pole [pəul] n. 极	9(A)
port [pɔ:t] n. 孔，口	4(A)
portion ['pɔ:ʃən] n. 部分	9(A)
positive ['pɔzətiv] n. 正极	7(A)
post-processor [pəust'prəusesə] n. 后处理程序	5(B)
power-driven ['pauə'drivn] a. 电动的	3(A)
precision [pri'siʒən] n. 精度	3(B)
pre-cut ['pri:'kʌt] v. 预切	2(B)
preferred ['pri'fə:d] a. 优先的，首选的	1(A)
preprogrammed [pri'prəugræmd] a. 预先编制好的	3(B)
preset ['pri:'set] v./n. 预设	5(B)
press [pres] n. 冲床	4(B)
principle ['prinsəpl] n. 原理	9(A)
procedure [prə'si:dʒə] n. 步骤	6(B)
proceed [prə'si:d] v. 继续	5(B)
process [prə'ses] n. 工艺	6(A)
processor ['prəusesə] n. 处理器	8(A)
profiling ['prəufailiŋ] n. 设置文件	5(B)
program ['prəugræm] n. 程序	
v. 编程	5(A)
programmable ['prəugræməbl] a. 可编程的	5(A)
programmer ['prəugræmə] n. 编程员	5(A)
programming ['prəugræmiŋ] n. 编程	8(B)

projection [prə'dʒekʃən] n. 投影	6(A)
propensity [prə'pensiti] n. 倾向性	1(B)
property ['prɔpəti] n. 属性	2(A)
proportion [prə'pɔːʃən] n. 比例	6(B)
proportional [prə'pɔːʃənl] a. 成比例的	7(B)
proportionately [prə'pɔːʃənitli] adv. 按比例地	8(B)
pulse [pʌls] n. 脉冲	8(B)
punch [pʌntʃ] n. 冲头，冲床	1(A)
purpose ['pəːpəs] n. 作用	5(A)
push [puʃ] v. 流出	7(B)
push-button [puʃ'bʌtn] n. 按钮	10(A)
quartz [kwɔːts] n. 石英	8(A)
quench [kwentʃ] v. （淬火）冷却	1(A)
radial ['reidjəl] a. 径向的	1(B)
rag [ræg] n. 地脚螺栓	1(B)
ram [ræm] n. 连杆，撞锤	2(B)
razor ['reizə] n. 剃刀	2(A)
readout ['riːdaut] n. 读数,读数器	1(A)
real-time ['riːəltaim] a. 实时的	8(A)
reamer ['riːmə] n. 铰刀	2(A)
recess [ri'ses] n. 凹窝处	1(A)
recessed [ri'sesd] a. 凹陷的	4(B)
recombine [riːkəm'bain] v. 重新组合	7(A)
recovery [ri'kʌvəri] n. 恢复	2(B)
Recrystallization [riː'kristəlazeiʃən] n. 再结晶	4(B)
rectangle ['rektæŋgl] n. 长方形	2(B)
rectifier ['rektifaiə] n. 整流器	7(A)
refine [ri'fain] v. 精加工	3(A)
register ['redʒistə] n. 寄存器	8(B)
relay ['riːlei] n. 继电器	7(A)
release [ri'liːs] v. 释放	10(B)
reliability [ri,laiə'biliti] n. 可靠性	10(A)
remote [ri'məut] a. 远程的	7(A)
removal [ri'muːvəl] n. 切削	3(A)
remove [ri'muːv] v. 拆除	6(A)
represent [,riːpri'zent] v. 表示，代表	6(A)
repulsion [ri'pʌlʃən] n. 排斥	9(A)
reservoir ['rezəvwɑː] n. 流体箱	4(A)
residual [ri'zidjuəl] a. 残余的	2(B)

resistance [ri'zistəns]　n. 电阻（值）　　　　　　　　　　　7(A)
resistant [ri'zistənt]　a. 抵抗的，阻止的　　　　　　　　　1(A)
resistor [ri'zistə]　n. 电阻器　　　　　　　　　　　　　　7(A)
resonator ['rezəneitə]　n. 谐振器　　　　　　　　　　　　8(A)
resulting [ri'zʌltiŋ]　a. 结果的，作为结果的　　　　　　　3(A)
retract [ri'trækt]　v. 拉回　　　　　　　　　　　　　　　4(A)
reverse [ri'və:s]　a. 反的，使反向　　　　　　　　　　　　5(A)
revolution [,revə'lu:ʃən]　n. 转数　　　　　　　　　　　　5(A)
rivet ['rivit]　n. 铆钉　　　　　　　　　　　　　　　　　2(A)
robustness [rə'bʌstnis]　n. 强度　　　　　　　　　　　　7(A)
rod [rɔd]　n. 杆　　　　　　　　　　　　　　　　　　　1(A)
rotary ['rəutəri]　a. 旋转的　　　　　　　　　　　　　　5(A)
rotate [rəu'teit]　v. 旋转　　　　　　　　　　　　　　　1(B)
rotation [rəu'teiʃən]　n. 旋转　　　　　　　　　　　　　9(B)
rotor ['rəutə]　n. 转子　　　　　　　　　　　　　　　　5(A)
roughing ['rʌfiŋ]　n. 粗加工　　　　　　　　　　　　　5(B)
route [ru:t]　n. 流道
　　　　　　　v. 按某线路传送　　　　　　　　　　　　　4(A)
rubber ['rʌbə]　n. 橡胶　　　　　　　　　　　　　　　　6(B)
rust [rʌst]　n. 锈　　　　　　　　　　　　　　　　　　1(A)
sandpaper ['sændpeɪpə(r)]　n. 砂纸　　　　　　　　　　1(B)
save [seiv]　v. 保存　　　　　　　　　　　　　　　　　5(B)
saw [sɔ:]　n. 锯　　　　　　　　　　　　　　　　　　　1(B)
scan [skæn]　v. 扫描　　　　　　　　　　　　　　　　10(A)
screw [skru:]　n. 螺钉　　　　　　　　　　　　　　　　1(A)
screwdriver ['skru:draivə]　n. 螺丝刀　　　　　　　　　2(A)
seal [si:l]　n. 密封套　　　　　　　　　　　　　　　　　4(A)
seal-in [si:lin]　a. 自锁的　　　　　　　　　　　　　　10(B)
secondary ['sekəndəri]　a. 次要的　　　　　　　　　　　3(A)
secure [si'kjuə]　v. 固定　　　　　　　　　　　　　　　3(B)
securely [s'kjuəli]　ad. 安全地　　　　　　　　　　　　1(A)
segment ['segmənt]　n. 段　　　　　　　　　　　　　　8(A)
self-sufficiency [,selfsə'fiʃəns]　n. 自供应（不用外接硬件）　8(A)
semiconductor ['semikən'dʌktə]　n. 半导体　　　　　　7(A)
sensor ['sensə]　n. 传感器　　　　　　　　　　　　　　8(A)
sequence ['si:kwəns]　n. 次序　　　　　　　　　　　　6(B)
sequencing ['si:kwənsiŋ]　n. 测序　　　　　　　　　　10(A)
sequential [si'kwinʃəl]　a. 顺序的　　　　　　　　　　　5(A)
serial ['siəriəl]　a. 串行的　　　　　　　　　　　　　　8(A)

| Glossary

series ['siəri:z]   n. 系列   6(B)
servomotor ['sə:vəu,məutə]   n. 伺服电机   5(A)
set [set]   n. 组   5(A)
setback ['setbæk]   n. 缩入距离   2(B)
shaft [ʃɑ:ft]   n. 传动轴   1(A)
shank [ʃæŋk]   n. 柄   2(A)
shape [ʃeip]   v. 刨削，使成形   3(A)
shear [ʃiə]   n. 剪切   2(A)
shearing ['ʃiəriŋ]   n. 切断加工   4(B)
sheet [ʃi:t]   n. 金属板   2(B)
shim [ʃim]   n. 垫片   1(B)
Siemens ['si:mənz]   n. 西门子   7(B)
silicon ['silikən]   n. 硅   2(A)
silvery ['silvəri]   a. 银色的，似银的   1(A)
simulation [,simju'leiʃən]   n. 模拟   5(B)
single-turn ['siŋgltə:n]   a. 单圈的   9(A)
size [saiz]   n. （图纸）号   6(B)
sketch [sketʃ]   n. 草图，素描   6(B)
slice [slais]   v. 切开，切成薄片   6(A)
slide [slaid]   n. 滑尺
　　　　　　 v. 滑动   1(A)
solder ['sɔldə]   v. 焊接   7(A)
smooth [smu:ð]   v. 消除   8(B)
solenoid ['səulinɔid]   n. 螺线管，电磁阀   10(A)
solid ['sɔlid]   a. 实体的
　　　　　 v. 焊接   6(A)
solid-state ['sɔlidsteit]   a. 固态的   7(A)
source [sɔ:s]   n. 电源   7(B)
spark [spɑ:k]   n. 火花   1(A)
spatial ['speiʃəl]   a. 空间的   6(A)
spawn [spɔ:n]   v. 促进   8(B)
speaker ['spi:kə]   n. 话筒   8(B)
special-purpose ['speʃəl'pə:pəs]   a. 专用的   10(A)
specification [,spesifi'keiʃən]   n. 技术规范，技术参数   3(A)
specify ['spesifai]   v. 限定   5(A)
spin [spin]   v. 旋转   8(B)
spindle ['spindl]   n. 伸缩爪（轴）   1(A)
spline [splain]   n. 曲线规   6(B)
spool [spu:l]   n. 线轴   4(A)

词条	页码
spring-back [spriŋbæk]　n. 回弹	2(B)
squeeze [skwi:z]　v. 挤压	4(B)
stack [stæk]　v. 堆	
n. 堆栈	1(B)
stackable ['stækəbl]　a. 叠加式的	4(A)
stainless ['steinlis]　a. 不生锈的	1(A)
stamping ['stæmpiŋ]　n. 冲压件	2(A)
statement ['steitmənt]　n. 语句	10(A)
stationary ['steiʃ(ə)nəri]　a. 静止的	2(B)
stator ['steitə]　n. 定子	5(A)
status ['steitəs]　n. 状态	10(B)
stepper ['stepə]　n. 步进电机	8(A)
stiffen ['stifn]　v. 硬化	7(A)
stock [stɔk]　n. 棒料，工料，原料	1(A)
straightedge ['streitedʒ]　n. 直尺	6(B)
strain-hardening [strein'hɑ:dəniŋ]　a. 应变硬化的	4(B)
strength [streŋθ]　n. 强度	9(A)
stress [stres]　n. 应力	2(B)
stretch [stretʃ]　v./n. 伸展	2(B)
strip [strip]　n. 金属片，条	1(A)
strobe [strəub]　n. 选通脉冲	8(B)
structure ['strʌktʃə]　v. 组成	10(A)
stubby ['stʌbi]　a. 粗而短的	1(A)
sub-discipline ['sʌb'disiplin]　n. 分支学科	6(B)
substitute ['sʌbstitju:t]　n. 替代品	1(B)
suspend [səs'pend]　v. 悬置	6(A)
swaging ['sweidʒiŋ]　n. 型锻	4(B)
swarf [swɔ:f]　n. 金属切屑	3(A)
switch [switʃ]　v. 切换	9(A)
swivel ['swivl]　v. 使旋转，使回旋	
n. 转轴	4(A)
symbol ['simbəl]　n. 符号	7(A)
synthetic [sin'θetic]　a. 合成的	1(B)
tailstock ['teilstɔk]　n. 尾座	1(A)
tank [tæŋk]　n. 水槽	4(A)
tap [tæp]　n. 螺丝攻	
v. 攻丝	2(A)
taper ['teipə]　n. 锥度	1(B)
tapered ['teipəd]　a. 锥形的	2(B)

| Glossary |

technician [tek'nɪʃ(ə)n]   n. 技术员   6(A)
tensile ['tensail]   a. 可拉伸的   2(A)
tension ['tenʃən]   n. 拉张，张力   2(B)
terminal ['tə:minl]   n. 端子，终端   7(B)
texture ['tekstʃə]   n. 纹理   3(A)
thermal ['θə:məl]   a. 热的   2(A)
thickness ['θiknis]   n. 厚度   2(B)
thread [θred]   n. 螺纹   1(B)
three-dimensional [θri:'dimenʃənəl]   a. 三维的   6(A)
thyristor [θai'ristə]   n. 晶闸管   7(A)
timer ['taimə]   n. 定时器   8(A)
timing ['taimiŋ]   n. 定时   10(A)
titanium [tai'teinjəm, ti]   n. 钛   2(B)
tolerance ['tɔlərəns]   n. 公差   4(B)
tooling ['tu:liŋ]   n. 工具，工装   3(B)
torque [tɔ:k]   v. 施以转动力
                 n. 转力矩，转矩   4(A)
toughness ['tʌfnis]   n. 韧性   2(A)
track [træk]   n. 轨道   3(B)
transfer [træns'fə:]   v./n. 传送   10(A)
transistor [træn'zistə]   n. 晶体管   7(A)
transmission [trænz'miʃən]   n. 传输，传送   4(A)
transmit [trænz'mit]   vt. 传输，转送   4(A)
treatment ['tri:tmənt]   n. 处理   10(A)
trial ['traiəl]   n. 试刀   1(B)
triangle ['traiæŋgl]   n. 三角板   6(B)
T-square [ti:-skweə]   n. 丁字尺   6(B)
tubing ['tju:biŋ]   n. 管材   2(A)
tune [tju:n]   v. 调整   5(B)
tungsten ['tʌŋstən]   n. 钨   2(A)
turret ['tʌrit]   n. 转塔   3(B)
tutorial [tju:'tɔ:riəl]   n. 教程   5(B)
typeface ['taipfeis]   n. 字型   6(A)
unbroken ['ʌn'brəukən]   a. 不断开的   7(B)
undercoat ['ʌndəkəut]   n. 底漆，底层镀   2(A)
undercut ['ʌndəkʌt]   n. 侧分型   4(B)
undergo [ˌʌndə'gəu]   v. 经受   2(B)
upper ['ʌpə]   a. 上部的   9(A)
upsetting [ʌp'setiŋ]   n. 镦粗加工   4(B)

单词	页码
utensil [ju(:)'tensl] n. 厨具	2(A)
vacuum ['vækjuəm] v. 用吸尘器吸 n.真空	1(A)
valve [vælv] n. 阀门	2(A)
variable ['vɛəriəbl] n. 变量	7(B)
verbally ['və:bəli] ad. 口头上	7(B)
version ['və:ʃən] n. 版本，型号	1(B)
vertical ['və:tikəl] a. 垂直式的	3(B)
vertically ['və:tikəli] ad. 沿垂直方向	2(A)
vibrate [vai'breit] v. 振动	8(B)
vibration [vai'breiʃən] n. 振动	4(A)
view [vju:] n. 视图，图	
v. 查看	5(B)
virtual ['və:tjuəl] a. 虚拟的	10(B)
vise [vais] n. 虎钳	1(B)
visually ['vizjuəli] ad. 可视地	6(B)
volatile ['vɔlətail] a. 不定的	8(B)
volt [vəult] n. 伏特	7(B)
washer ['wɔʃə] n. 垫圈	9(B)
wear [wεə] n. 磨损	2(A)
weld [weld] n. 焊接	
v. 焊接	2(A)
winding ['waindiŋ] n. 绕组	5(A)
wire ['waiə] v. 连接	8(B)
withdraw [wið'drɔ:] v. 取出	4(B)
wobble ['wɔbl] v. 摆动	1(A)
word [wə:d] n. 指令字	5(B)
workstation ['wɜ:ksteiʃ(ə)n] n. 工作站	6(B)
worm [wə:m] n. 蜗杆	2(A)
wrench [rentʃ] n. 扳手	1(B)
yoke [jəuk] n. 轭	9(B)
zinc [ziŋk] n. 锌	2(A)

Phrases

a collection of	很多，一批	3 (A)
a good collection of	大量的	1 (A)
a large range of	大量的，范围的	7 (A)
a sequence of	一系列的	2 (B)
a variety of	多种	3 (B)
absolute scale	绝对温标	4 (B)
academic discipline	学科	6 (A)
acts as	作为	7 (A)
address bus	地址总线	8 (A)
air pressure	气压	4 (B)
Alternating Current (AC)	交流	8 (A)
Analog-to-digital Converter (DAC)	模数转换器	8 (A)
Three Dimensions (3D)	三维	6 (B)
as long as	只要	7 (B)
Automatic Tool Changer (ATC)	自动换刀装置	5 (A)
axis of motion	运动轴	5 (A)
axis of rotation	旋转轴	3 (A)
ball bearing	滚珠轴承	5 (A)
bend angle	弯曲角	2 (B)
band-saw	带锯	1 (B)
base address	基址	8 (B)
be available for	用于	10 (A)
be based on	基于	1 (B)
be classified as	分类为	2 (A)
be combined with	被动与……结合	3 (A)
be comprised of	由……组成	10 (A)
be dedicated to	专门用于	10 (A)
be designed to	用于	8 (B)
be directly proportional to	与……成正比	7 (B)
be handy for	便于	1 (A)
be in accordance with	依据	6 (A)
be in series with	与……串联	10 (B)
be inversely proportional to	与……成反比	7 (B)

英文	中文	章节
be known as	被称为	7 (B)
be made up of	由……组成	7 (A)
be parallel to	与……平行	3 (A)
be proportional to	与……成比例	7 (B)
be referred to as	被称为	3 (B)
be required to	用于	10 (A)
be square with	与……垂直	1 (A)
bend allowance	弯曲余量	2 (B)
bend axis	弯曲轴线	2 (B)
bend deduction	折弯补偿	2 (B)
bend length	弯曲长度	2 (B)
bend line	弯曲线	2 (B)
bend radius	弯曲半径	2 (B)
bevel angle	斜角	2 (B)
blank holder	压料板	2 (B)
block diagram	框图	7 (B)
bottom die	底模	4 (B)
boundary line	分界线	2 (B)
bow out	跳离	1 (A)
by contrast	相反	1 (B)
by hand	用手工	1 (B)
by means of	通过	3 (A)
by no means	绝不	1 (B)
carbon steel	碳素钢	1 (A)
case seal	密封垫	4 (A)
cast iron	铸铁	1 (A)
center drill	中心钻	1 (A)
Central Processing Unit (CPU)	中央处理器	8 (A)
charge pump	补油泵	4 (A)
chip brush	清屑刷	1 (A)
circular saw	圆盘锯	1 (B)
clock generator	时钟发生器	8 (A)
closed center circuit	闭式回路	4 (A)
closed-die forging	闭式模锻	4 (B)
coating unit	涂料机	4 (B)
cold forging	冷锻	4 (B)
compressive force	挤压力	4 (B)
Computer Aided Design (CAD)	计算机辅助设计	6 (B)
Computer Numerical Control (CNC)	计算机数字控制	5 (A)

Computer Numerical Control (CNC)　计算机数控	3 (B)
contrasting color　对比色	1 (A)
contribute to　对……起作用	8 (B)
control cabinet　控制柜	10 (A)
control valve　控制阀	4 (A)
conventional machine　常规机床	5 (A)
counter bore　沉孔	5 (A)
cross section　横截面	2 (B)
cross slide　横刀架	3 (B)
current source　电流源	7 (B)
curved surface　曲面	2 (B)
cutaway drawing　剖视图	6 (A)
Data Block (DB)　数据模块	10 (A)
data bus　数据总线	8 (A)
define…as…　把……定义为……	7 (A)
depth of cut　切削深度	3 (B)
descriptive geometry　画法几何	6 (A)
design collaboration　协同设计	6 (B)
dial caliper　带表卡尺	1 (A)
die cavity　型腔	2 (B)
die forging　模锻	4 (B)
digital caliper　数显卡尺	1 (A)
dimensional tolerance　尺寸公差	4 (B)
direct coupling　直接联轴器	5 (A)
Direct Current (DC)　直流	7 (B)
directional valve　换向阀	4 (A)
divide into　分成	9 (A)
double throw switch　双向开关	7 (A)
drafting paper　绘图纸	6 (B)
drafting table　制图台	6 (B)
draughts person　制图员	6 (B)
draw reduction　深拉补充	2 (B)
drift pin　冲头	1 (B)
drill press　钻床	3 (A)
drill rod　钻杆	1 (A)
drive gear　主动齿轮，驱动齿轮	4 (A)
drive shaft　驱动轴	4 (A)
drop hammer die　落锤模	2 (A)
drum sequencing　鼓测序	10 (A)

elastic band　松紧带　9 (A)
Electrical Discharge Machining (EDM)　放电加工　3 (A)
Electrically Erasable Programmable Read Only Memory (EEPROM)　电可擦可编程序只读存储器　8 (A)
Electromotive Force (EMF)　电动势，电磁力　7 (B)
energy consumption　能耗　7 (A)
energy gap　能隙　7 (A)
engineered item　工程部件　6 (A)
engineering drawing　机械制图　6 (A)
Erasable Programmable Read Only Memory (EPROM)　可擦可编程序只读存储器　8 (A)
excel at　擅长　1 (A)
exploded view drawing　分解图　6 (A)
fall into　分类为　3 (A)
feeder automation unit　自动进料机　4 (B)
feeler gage　测隙规　1 (A)
ferrous metal　黑色金属　2 (A)
fine finishing　精加工　3 (A)
finish machining　精加工　4 (B)
fixed volume pump　定量泵　4 (A)
flange length　卷边长度　2 (B)
flash memory　闪存　8 (A)
Fleming's left hand rule　弗莱明左手规则　9 (A)
float center directional valve　Y型中位机能电磁换向阀　4 (A)
follower rest　跟刀架　1 (A)
forging die　锻模　4 (B)
forging machine　锻造机　4 (B)
form die　成型模　2 (B)
freehand drawing　徒手图　6 (B)
French curve　曲线板　6 (B)
front end bell　前端罩　5 (A)
front view　正视图　10 (B)
Function Block (FB)　功能模块　10 (A)
geometric feature　几何特征　6 (A)
give rise to　产生　10 (A)
go through　完成　10 (A)
grain refinement　晶粒细化　4 (B)
gripping jaw　鄂形夹爪　2 (B)
hammer forging　锤锻　4 (B)
hand wheel　手轮　3 (B)

Harvard architecture 哈佛结构	8 (A)
headstock assembly 主轴箱	3 (B)
heat treatment 热处理	4 (B)
heavy machinery 重型机械	4 (B)
High Speed Steel (HSS) 高速钢	1 (B)
holding force 夹持力	2 (B)
home appliance 家用电器	8 (A)
hot forging 热锻	4 (B)
hydraulic cylinder 液压缸	4 (A)
hydraulic press 液压机	4 (B)
hydraulic pump 液压泵	4 (A)
hydraulic tank 油槽	4 (A)
hydraulic transmission 液压传动	4 (A)
hydrostatic transmission 液压静力传动	4 (A)
idler gear 从动齿轮，空转齿轮	4 (A)
impression-die forging 飞边模锻	4 (B)
in light of 鉴于	10 (A)
in practice 在实际中	7 (B)
in terms of 在……方面	6 (B)
in the form of 以……形式	7 (A)
industrial arts 工艺美术	6 (B)
Insulating coating 绝缘涂层	9 (B)
intended purpose 预期作用	1 (A)
Jacobs taper 贾克布锥度	1 (B)
ladder diagram 梯图	10 (A)
LCD (Liquid Crystal Display) 液晶显示器	1 (A)
lead screw 丝杠	2 (A)
lead wire 引线	2 (A)
light oil 轻质油	1 (B)
light source 光源	7 (A)
Light-emitting Diode (LED) 发光二极管	7 (A)
limit switch 限位开关	10 (A)
Line Print Terminal (LPT) 打印终端	8 (B)
line of magnetic force 磁力线	9 (A)
line style 线型	6 (A)
line thickness 线的粗细	6 (A)
live center 活动顶尖	1 (B)
machine cycle 加工周期	5 (A)
machine tool 机床	5 (A)

machining center 加工中心	5 (A)
machining property 加工性能	1 (A)
magnetic field 磁场	7 (A)
magnetic flux 磁通	9 (A)
manual drafting 手工制图	6 (B)
manufacturing process 生产过程	5 (A)
mechanical property 机械特性	4 (B)
Medium Density Fiberboard (MDF) 中密度纤维板	1 (B)
melting temperature 熔化温度、熔点	4 (B)
mild steel 低碳钢	2 (B)
more recent 最近	3 (A)
Morse Taper 莫氏锥度	1 (B)
mounting flange 安装法兰	4 (A)
non-ferrous metal 有色金属	2 (A)
non-normal contact 常闭触点	10 (B)
non-volatile memory 永久性存储器	8 (A)
normally-closed contact 常闭触点	10 (B)
normally-open contact 常开触点	10 (B)
NPN transistor NPN 型晶体管	8 (B)
Ohm's law 欧姆定律	7 (B)
oil tank 油箱	4 (A)
on the basis of 基于	10 (A)
open center circuit 开式回路	4 (A)
open-die forging 开式模锻	4 (B)
Organization Block (OB) 组织模块	10 (A)
outside mold line 外模线	2 (B)
packing grease 密封润滑脂	1 (A)
parallel line 平行线	6 (B)
parallel rule 平行规	6 (B)
parts feeder 拾取定向料斗	4 (B)
patent drawing 专利图	6 (A)
peeling machine 装料机	4 (B)
peripheral equipment 外设	8 (A)
permanent magnet 永久磁铁	5 (A)
Personal Computer (PC) 个人计算机	8 (A)
physical form 实物形式	6 (A)
plug in 插上插头接通电源	8 (B)
positional accuracy 位置精度	5 (A)
pounds per square inch (psi) 磅/平方英寸	4 (A)

power supply	电源	7 (B)
powered tool	电动工具	3 (B)
press forging	压锻	4 (B)
press table	下压板	2 (B)
pressure port	压油孔，压力孔	4 (A)
pressurized hydraulic fluid	压力流体	4 (A)
primary motion	主运动	3 (A)
Printed Circuit Board (PCB)	印刷电路板	7 (A)
prior to	在……之前	4 (B)
Program Block (PB)	程序模块	10 (A)
Programmable Logic Controller (PLC)	可编程逻辑控制器	10 (A)
projected view	投影视图	6 (B)
Pulse-Width Modulation (PWM)	脉冲宽度调制	8 (B)
pumping station	泵站	10 (A)
PWM generator	脉冲宽度调制器	8 (B)
Quick Change Tool Post (QCTP)	快换刀架	1 (B)
radial arm saw	手拉锯	1 (B)
Radom Access Memory (RAM)	随机存储器	8 (A)
raw material	原材料	1 (A)
Read Only Memory (ROM)	只读存储器	8 (A)
real-time clock	实时时钟	8 (A)
rear end bell	后端罩	5 (A)
relief valve	电磁溢流阀	4 (A)
residual stress	残余应力	2 (B)
rest on	位于	3 (B)
Revolutions Per Minute (RPM)	每分钟转数	5 (A)
right angle	直角	6 (B)
right-handed screw rule	右手螺旋定则	9 (A)
rock drill	凿岩机	2 (A)
roll forging	辊锻	4 (B)
rotor laminations	转子铁芯	5 (A)
saber saw	军刀形电动手锯	1 (B)
schematic diagram	n. 框图	10 (B)
seal-in contact	自锁触点	10 (B)
secondary motion	次运动	3 (A)
self oscillator	自激振荡器	8 (B)
Sequence Block (SB)	顺序模块	10 (A)
sequential order	顺序次序	5 (A)
serial port	串行口	8 (A)

英文	中文	位置
serve as	作为	10 (B)
set out	做出	3 (A)
set screw	紧定螺钉	2 (A)
set square	三角板	6 (B)
silicon controlled rectifier	可控硅整流器	7 (A)
Single Chip Microprocessor (SCM)	单片机	8 (A)
socket head screw	凹头螺钉	1 (B)
spark plug	火花塞	1 (A)
spatial ordering	空间排序	6 (A)
specialize in	专门从事	3 (A)
spindle speed	轴转速	5 (A)
stack pointer	堆栈指针	8 (B)
stainless steel	不锈钢	2 (A)
standard practice	常规做法	1 (A)
stand for	代表	5 (A)
steady rest	固定支架	1 (A)
stepper motor	步进发电机	5 (A)
straight line	直线	9 (A)
straight side	直边	6 (B)
stretch press	拉伸机	2 (B)
suction port	进油孔，吸入孔	4 (A)
switching device	开关装置	10 (A)
table saw	台锯	1 (B)
tailstock assembly	尾架	3 (B)
take up	占用	10 (A)
technical drawing	技术制图	6 (A)
technical illustration	技术图解	6 (A)
technical pen	针笔	6 (B)
tend to	易于	1 (B)
tensile force	拉力	2 (B)
tensile strength	抗张强度	2 (A)
text size	文本长度	6 (A)
T-handle Metric Hex Wrench Set	T型手柄公制成套六角扳手	1 (B)
threading tool	螺纹车刀	1 (B)
three-dimensional model	三维模型	6 (A)
time-delay relay	时滞继电器	10 (A)
timing circuit	定时电路	7 (A)
tool bit	刀头	1 (B)
Tool Blank	刀头	1 (B)

| Phrases

tool magazine　刀库　　　　　　　　　　　　　　　　　　　　　　　　　5 (A)
tool post　刀座　　　　　　　　　　　　　　　　　　　　　　　　　　3 (B)
top die　上模　　　　　　　　　　　　　　　　　　　　　　　　　　　4 (B)
TRIAC　双向可控硅　　　　　　　　　　　　　　　　　　　　　　　10 (B)
Two Dimensions (2D)　二维　　　　　　　　　　　　　　　　　　　　6 (B)
Universal Asynchronous Receiver/ Transmitter (UART)　异步串行接口　　8 (A)
Universal Serial Bus (USB)　通用串行总线　　　　　　　　　　　　　　8 (A)
up to　多达　　　　　　　　　　　　　　　　　　　　　　　　　　　　2 (A)
user program　用户程序　　　　　　　　　　　　　　　　　　　　　　10 (A)
vary with　随着……的变化而变化　　　　　　　　　　　　　　　　　7 (B)
view projection　视图投影　　　　　　　　　　　　　　　　　　　　　6 (A)
visual appearance　视觉外观　　　　　　　　　　　　　　　　　　　　6 (A)
voltage source　电压源　　　　　　　　　　　　　　　　　　　　　　7 (B)
Walden Specialties　瓦尔登专卖店　　　　　　　　　　　　　　　　　1 (B)
warm forging　温锻　　　　　　　　　　　　　　　　　　　　　　　　4 (B)
white lithium grease　白色锂基润滑脂　　　　　　　　　　　　　　　　1 (B)

第1单元 机械技术基础

课文A 车床附件（I）

铜质圆棒料

与铝或钢相比，铜虽然有点贵，但它却是一种用于加工的优良材料。它能给将要展示的工件的对比色增光添彩。国内工厂最常使用的铜合金牌号是360铜（见图1-1）。

中心钻

中心钻是刚性的、粗而短的小型钻头，它用于在工件端面开孔。如果你在工件上钻孔时不使用中心钻，钻头很可能摆动而偏离中心，并且会在工件上钻出斜孔（见图1-2）。

图1-1 铜质圆棒料

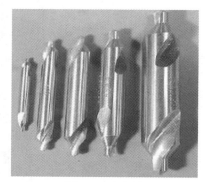
图1-2 中心钻

标准的钻削做法是首先在工件一端车端面，然后用中心钻开个孔，接下来用标准钻头钻孔钻到所需的深度。

清屑刷

清屑刷就是廉价的油漆刷子，在车间里到处都在使用。我们不确定清屑刷名字的由来，但我们认为它最初与清除铁屑无关。不过现在清屑是这些刷子所擅长的用途之一。用它们清洁新车床和新铣床上的密封润滑脂也很理想。顺便说一句，经验丰富的机械师们会告诉你要用刷子而非压缩空气去清洁在机床上的铁屑，因为压缩空气会驱使铁屑进入机床的缝隙深处。使用清屑刷和工业吸尘器清除铁屑是首选的方法（见图1-3）。

数显卡尺

数显卡尺就像带表卡尺一样，被用来实现千分之一英寸的内外表面精确测量，不过它可直接通过液晶显示器进行数字读数。在游标卡尺上，您可以首先从滑尺上读出十分之几英寸的大尺寸，同时在心里加上刻度尺上千分之几英寸的小尺寸。一段时间以后，这个过程就变成一种自然习惯，但仍然会带来出错的机会。数显卡尺会在显示屏上读出全部尺寸，所以是相当方便，只要你事先把它归零，就不会出错。你也可以按照需要切换米制和英制的模式（见图1-4）。

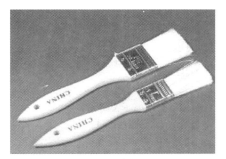

图1-3　清屑刷

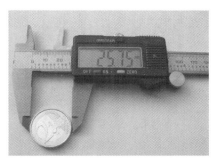

图1-4　数显卡尺

钻杆

钻杆是闪亮的银色钢合金，具有良好的加工性能。不像其他原材料，公称直径的变化达+0.010或更多，钻杆表面经过磨削，公称直径尺寸误差大约在0.001毫米以内。虽然钻杆不归入不锈钢，但至少和普通碳素钢相比，它具有较好的防锈性能。这一点使它非常适用于传动轴和车轴。

钻杆的一个有用特性是它能容易地通过加到红热状态后，在油或水中急冷而淬火变硬，这样处理后，金属会足够硬，可用作如冲头这样的工具（见图1-5）。

成组钻头

钻孔是在车床上最常见的操作之一，因此你需要大量的优质钻头。买车床时，别忘了订购尾座夹头和刀杆来装夹钻头。

质量差的钻头很容易碰到，但购买它们纯属浪

图1-5　钻杆

费金钱。这并不是说你必须购买顶级的工业钻头只通过看外表来区分钻头的好坏并不容易，当然，如果你只有网站上或目录中的图片，那就更难区分了。不过从总体而言，不要买价格最低的成组钻头（见图1-6）。

花盘

花盘是一个很方便的附件，用于车削不易固定在夹具上的形状奇特的工件。尽管自己亲

自用钢或铝制作花盘并不困难，但是以购买的价格来看，付出这样的努力就不值得了。

在把花盘安装于主轴后，常规做法是需进行一次端面的轻切削，以确保花盘表面与车床主轴垂直。呼入铸铁粉末是非常有害的，所以在操作期间，强烈推荐戴防尘口罩直到你用吸尘器吸净所产生的粉末。

当使用花盘时，要始终确保工件安全夹紧，同时，如果需要可通过放置偏心金属块使系统保持平衡（见图1-7）。

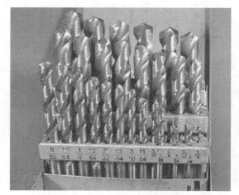

图1-6　钻具组件

图1-7　面板花盘

测隙规

测隙规是几组具有准确厚度的薄钢片，厚度通常为0.001～0.040英寸。一组测隙规中通常有20～40个独立的钢片，它们通过一个穿过每个测隙规末端孔的螺栓连在一起。每个测隙规上面都标有厚度值，下面是标有英寸数和毫米数的样例。当不使用时，所有钢片都应折入手柄中防止弯曲。

它们的预期作用是测量两个面之间的间距，如火花塞的电极板之间的距离（见图1-8）。

跟刀架

跟刀架类似于固定支架，但是它位于刀具后面，被固定在溜板上，并与溜板同时移动，对工件起到移动支撑作用。加工刚性小的工件时，使用起来很方便，否则这些工件就会产生让刀现象。如果你以前不清楚溜板左边的两个螺丝孔起什么作用的话，现在你就知道了，它们是跟刀架的安装孔（见图1-9）。

图1-8　测隙规

图1-9　跟刀架

课文 B　车床附件（Ⅱ）

活动顶尖

尾架上所使用的顶尖是用来顶住相对较长而且灵活工件的端面。"死"顶尖的弊端是顶尖不能旋转，而死顶尖所顶住的工件可以旋转，这就会产生摩擦和可能造成过热。相反，"活"顶尖的尖端可自如地在轴承中旋转，而且可与工件共同旋转，大大减少了摩擦（见图1-10）。

图 1-10　活动顶尖

中密度纤维板

中密度纤维板并非是附件，但它是一种使用方便、价格低廉、重量很轻且易于加工的材料，在车间里使用广泛。它是一种合成材料，但具有不同的属性。它的厚度至少可达到 3/4 英寸，两个面都是光滑的，可以很简单地用军刀形电动手锯、圆盘锯、带锯、台锯或手拉锯进行加工。它的一个不足之处是易于吸收水分，但可通过表面喷漆大大地减小这种倾向性（见图1-11）。

铣刀夹具

铣刀夹具是小型车床的标准附件。在它出现之前，许多车床所有者根据 Varmint A1 公司所制造的铣刀夹具制作自己的夹具。Varmint A1 公司的铣刀夹具使用标准的铣刀虎钳固定工件，而图 1-12 所示的夹具使用了凹头螺钉将工件固定在夹具之间。该夹具绝不是小型铣床的替代品，但它是一种可以在一定限度内增加您的小型车床的铣削性能的廉价方式，同时也可节省铣削时间。

图 1-11　中密度纤维板

图 1-12　铣刀夹具

T 型手柄公制成套六角扳手

这些扳手并不是非常重要，但它们非常方便而且便宜。如果你不使用它们，就会给自己带来不便。后来又尝试这些成套六角扳手可用于几乎所有的车床凹头螺钉上（见图 1-13）。

尾座夹具和刀轴

钻孔是车床的一种基本操作，你需要使用一个夹具和一个 2 号大小莫氏锥度的刀轴来做这件事。刀轴的一端要有螺纹或贾克布锥度可与夹具匹配，另一端要有 2 号大小莫氏锥度可与尾座汽缸匹配。

为了能够移动刀轴，需把夹具置于台钳中，夹具的开口距离要比刀轴的直径大一点（不要夹住刀轴）。打开夹具后只需用一块短圆料或冲头就可将刀轴从夹具后部顶出。只需用锤子轻轻地敲击就能到这一点。把地脚螺栓定位于刀轴下面并固定它是个很好的办法，目的是不让它落到地面而发出声音（见图 1-14）。

图 1-13　T 型手柄公制成套六角扳手

图 1-14　尾座夹具和刀轴

刀头

建议你磨削自己的刀头时，要有一套预先磨好的高速钢刀头。它们不仅能让你立即开始加工，而且当你在磨刀头时可以得到你想要的刀样形状。这里有一套瓦尔登专卖店的特好刀头，它包括 60 个很难用手工来磨削的螺纹车刀（见图 1-15）。

快换刀架

快换刀架（QCTP）是一个我们强烈推荐使用的附件，只要预算允许就立即买一个吧。为什么呢？因为它能帮你节省大量时间并减轻你的压力，有了它你不再需要把垫片堆在刀头下使其与车床的中线高度匹配。由于使用了快换刀架，每个刀具都有它自己的专用刀架。每个刀架都可进行锁紧高度调整。经过几次快速试刀和调整后，你就能把刀具高度锁紧在合适的位置。除非刀尖磨损了千分之几而变得锐利了，否则无需调整刀架（见图 1-16）。

图 1-15　刀头

图 1-16　快换刀架

湿/干砂纸

湿/干砂纸非常易于增加车床金属加工件的抛光度。在此应用中通常使用干砂纸，但在抛光底部平面时通常使用湿砂纸，它会使得铣削加工金属件的抛光度更好（见图1-17）。

白色锂基润滑脂

白色锂基润滑脂对于润滑车床上的几乎任何运动部件都有用（滑道上使用轻质油，而非润滑脂）（见图1-18）。

图1-17　湿/干砂纸

图1-18　白色锂基润滑脂

第2单元　金属材料和金属成形

课文A　金属材料

几乎75%的元件都是金属。金属在电子设备中用作导线，在炊具中用作锅碗瓢盆，因为它们有良好的导电导热性能。大多数金属具有可塑性和延展性，而且通常比其他元素的物质要重。两种或更多种的金属可以形成合金材料，这些合金材料具有纯金属没有的属性。

金属可分为黑色金属或有色金属。黑色金属含铁，有色金属不含铁。所有的黑色金属都具有磁性但耐蚀性较差，而有色金属通常不具有磁性，却有较强的耐蚀性。最常见的黑色金属和有色金属如图2-1～图2-10所示。

黑色金属

（1）

材料名称：低碳素钢（见图2-1）

成分：含碳量最多0.30%

属性：成形性好、焊接能力强、成本低

用途：含碳量0.1%～0.2%：链条、冲压件、铆钉、钉子、电线、管等

　　　含碳量0.2%～0.3%：机床和结构件

（2）

材料名称：中碳素钢（见图2-2）

成分：含碳量0.30%～0.80%

属性：综合性能好、可成形性一般

用途：含碳量0.3%～0.4%：丝杠、齿轮、蜗杆、主轴、传动轴、机床零件

含碳量0.4%～0.5%：曲轴、齿轮、轮轴、心轴、刀柄、热处理过的机器零件

含碳量0.6%～0.8%：落锤模、紧定螺钉、螺丝刀、刀杆

含碳量0.7%～0.8%：韧性和硬质钢、铁砧面板、带锯、锤子、扳手、电缆心线

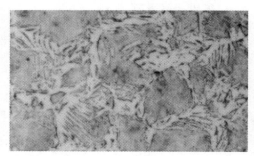

图2-1 低碳素钢

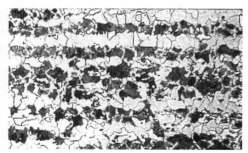

图2-2 中碳素钢

（3）

材料名称：高碳素钢（见图2-3）

成分：含碳量0.80%～2.0%

属性：韧性差、成形性和焊接能力弱、硬度和耐磨性高

用途：含碳量0.8%～0.9%：金属冲模、凿岩机、剪切刀片、冷凿、铆钉用具和许多手工工具

含碳量0.9%～1.0%：用于高硬度和高抗张强度的零件，如弹簧、刀具

含碳量1.0%～1.2%：钻头、螺丝攻、铣刀、刀具、冷加工模具、木工工具

含碳量1.2%～1.3%：锉刀、铰刀、刀具、切木和切铜刀具

含碳量1.3%～1.4%：用作如剃刀、锯等所需的锋利刀刃以及用于耐磨性很重要的地方

（4）

材料名称：不锈钢（见图2-4）

成分：不锈钢是一类耐腐蚀钢。它们含有至少10.5%的铬元素。合金中的铬形成了具有自我愈合能力且具有防护能力的清晰氧化层。这种氧化层使得不锈钢具有耐腐蚀性。

属性：耐腐蚀性强、外观好、机械性能好

用途：含铬量11.5%：广泛用于炊具、餐具、厨具。也用于五金用品、工业设备、建筑结构、汽车和航空航天业

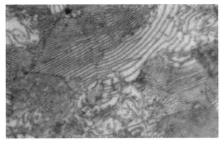
图 2-3　高碳素钢

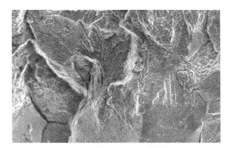
图 2-4　不锈钢

有色金属

（5）

材料名称：铝、铝合金（见图 2-5）
成分：纯金属易与少量铜、锰、硅、镁以及其他元素形成合金
属性：密度低、导电性良好（约为铜的 60%）、无磁性、不燃性、有韧性、可塑性、耐腐蚀、易成形、加工和铸造
用途：窗框、飞机零部件、汽车部件、厨具

（6）

材料名称：黄铜（见图 2-6）
成分：是铜和锌的合金，铜和锌的通常比率为 65%～35%
属性：硬度适当，铸造、成形和加工良好，导电性和声学性能良好
用途：电器配件、阀门、锻件、饰品、乐器

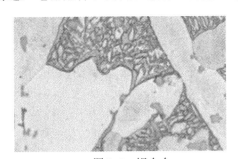
图 2-5　铝合金

图 2-6　黄铜

（7）

材料名称：铜（见图 2-7）
成分：纯金属
属性：韧性、导热性和热导电性极好
用途：电线、管材、壶、碗、管道、印制电路板

（8）

材料名称：镁/镁合金（见图 2-8）
成分：纯金属可与铝、铅、锌和其他有色金属形成合金；与铝形成合金可提高机械、制

造和焊接性能

属性：最轻的金属材料（密度大约为铝密度的2/3）、坚硬、最常用的可加工的金属、良好的耐腐蚀性、易于铸造

用途：汽车、便携式电子产品、家用电器、电动工具、体育用品配件、航空设备

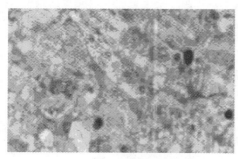

图2-7　铜

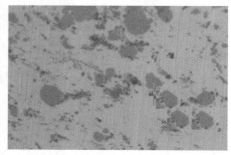

图2-8　镁合金

（9）

材料名称：镍/镍合金（见图2-9）

成分：纯金属非常容易与大量其他元素，主要是铬、钼、钨形成合金

属性：非常好的耐腐蚀性（合金的耐腐蚀性超过不锈钢）、良好的耐高温和机械特性、相当好的导热性和导电性

用途：用作装饰性镀铬中的底层电镀以提高耐腐蚀性，还用于电子引线、电池组件、易腐蚀环境下的换热器

（10）

材料名称：锌/锌合金（见图2-10）

成分：纯金用来与其他金属形成大量合金。与少量铜、铝、镁形成的主要合金——锌合金在压铸件中非常有用。

属性：极好的耐腐蚀性、重量轻、导电性适当

用途：主要用于钢铁镀锌（超过50%的锌用于钢铁镀锌）；因为重量轻，所以被广泛用于汽车领域。

图2-9　镍合金

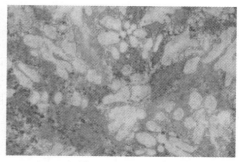

图2-10　锌合金

课文 B　金属板成形

弯曲成形

弯曲成形是把力施于金属板的成形过程,可使金属板弯曲成一定的角度,并形成期望的形状。弯曲操作可使金属板沿某一轴线变形,而执行一系列不同的操作就可形成复杂的零件。弯曲零件可小到支架,大到 20 英尺长的外壳和底盘。弯曲成形件的不同特性参数如下图所示(见图 2-11)。

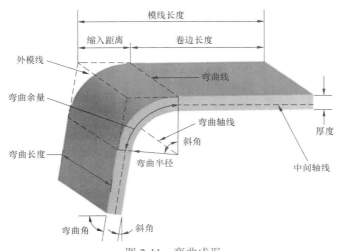

图 2-11　弯曲成形

弯曲的动作使金属板产生了拉张和压缩。金属板外部受到拉张并产生伸展,而内部经受到压缩而缩短。中性轴线是金属板的分界线,中性轴线上既无拉张力也无压缩力存在,因此,这个轴线的长度保持恒定。金属板内外表面长度的变化与最初平板长度有关,由两个参数决定:弯曲余量和折弯补偿(见图 2-12)。

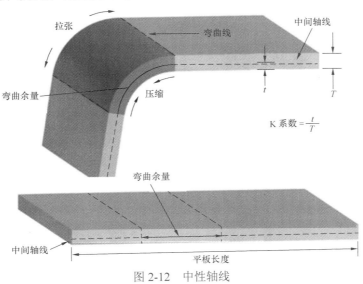

图 2-12　中性轴线

当弯曲金属板时，金属本身的残余应力会使得板在弯曲操作后轻微地回弹。正因为存在弹性恢复，所以有必要将金属板过度弯曲到某个精确的量以达到期望的弯曲半径和弯曲角度。最终的弯曲半径要比最初形成的大些，而最终的弯曲角要比最初形成的小些。最终弯曲角与最初弯曲角的比率被定义为回弹系数 Ks。决定回弹量的因素很多，如材料自身性质、弯曲操作方式、最初的弯曲角和弯曲半径（见图 2-13）。

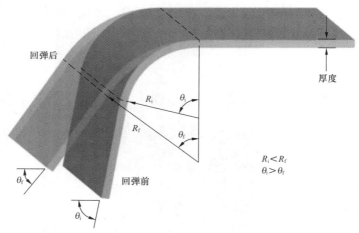

图 2-13　回弹

深拉成形

深拉成形是使金属板伸展为期望零件形状的金属成形过程。用外力将工具下压至金属板的型腔中产生期望的形状。施加在金属板上的拉力会使其塑性变形成杯状的元件。已深拉成形的零件以深度为特性，此深度值等于零件的大半个直径。这些零件有各种各样的横截面，如平面、锥面及曲面。易延展金属的深拉成形效果最明显，如铝、黄铜、铜以及低碳钢（见图 2-14）。

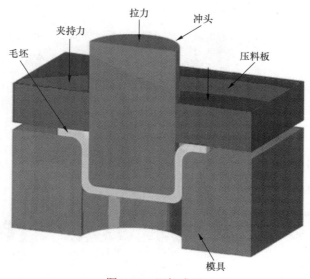

图 2-14　深拉成形

深拉成形过程需要有毛坯、压料板、冲头和模具。毛坯是一块金属板，通常是圆盘形或长方形，需从原料中预切下来并深拉成零件。

深拉零件的过程有时会出现在一系列的操作中，被称为深拉补充。在每个步骤中，冲头将材料压入不同的模具，同时每次都会将其拉伸到更大的深度。在零件被完全深拉后，冲头和压料板将会抬高，零件将会从模具中移出（见图2-15）。

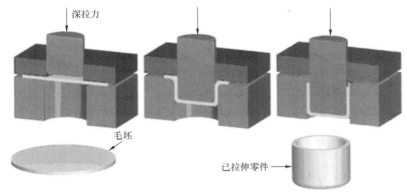

图2-15 深拉过程

拉伸成形

拉伸成形是金属成形过程，可使金属板拉伸并在模具上方弯曲成零件。拉伸成形操作在压板上执行，沿拉伸机的边缘，颚形夹爪把金属板安全地夹住（见图2-16）。

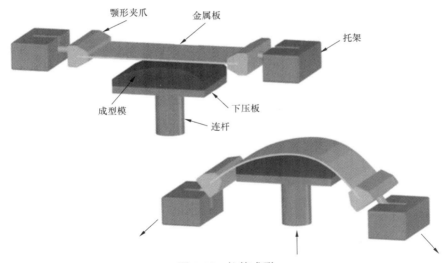

图2-16 拉伸成形

拉伸成形的零件通常很大，具有很大的弯曲半径。所拉伸的形状多种多样，或是简单的曲面或是复杂的不规则形状的横截面。拉伸成形加工后的成形零件准确度高、表面光滑。易延展的材料更适合拉伸成形加工，普遍使用的是铝、钢和钛。典型的拉伸成形后的零件是大型的曲面板，如汽车车门板和飞机机翼板。

最常见的拉伸机是垂直方向的，成型模位于下压板上，下压板可通过液压连杆上升至金属板。当成型模被驱动至金属板时，由于金属板的边缘被紧紧地夹在拉伸机的边缘处，所以随着张力的增加，金属板塑性变形为新的形状。水平方向的拉伸机将成型模固定在静止下压板的侧面，同时颚形夹爪会沿水平方向按成型模形状拉伸金属板。

第 3 单元　加工操作与车削机床

课文 A　加 工 操 作

常规加工是最重要的切除材料的方法之一，它包括很多材料加工工艺过程，使用多种电动机床：如车床、铣床和钻床等，这些机床使用锋利的刀具对材料进行机械加工，以获得期望的几何形状。加工是制造几乎所有金属制品过程的一部分。对其他材料进行加工也并不罕见。专门从事加工的人被称为机工。加工也可以是一种爱好。所从事加工的房间、建筑或公司也被叫作车间。大部分现代加工用计算机来控制，采用计算机数字控制系统（CNC）。

加工操作

三个主要的加工过程可分为车削、钻削、铣削。其他类的操作还包括牛头刨削、龙门刨削、镗孔、拉削、锯削（见图 3-1）。

车削操作是工件相对刀具旋转作为金属切削主运动的操作。车床是用于车削的主要机床。

铣削操作是刀具旋转带动刀刃相对工件运动的操作。铣床是用于铣削的主要机床。

钻削操作用于钻孔或孔的精加工，它通过钻头旋转来进行切削，钻头与工件接触那一端带有切削刃。钻削操作主要在钻床上完成，但是也经常在车床或铣床上完成。

最近，先进的加工技术还包括使用电火花加工（EDM）、电解加工、激光加工或水切割等来使工件成形。加工过程中要注意工件的许多细节，以满足工程图或蓝图中的技术规范。

各种类型的加工操作

有许多类型的加工操作，每一种都能生成一定的几何形状和表面纹理。

在车削加工中，具有单刀刃的刀具被用来从旋转的工件上切削材料以产生圆柱形。在车削中，旋转的工件实现速度运动，刀具慢慢沿着与工件旋转轴平行的方向移动实现进给运动（见图 3-2）。

图 3-1　加工操作

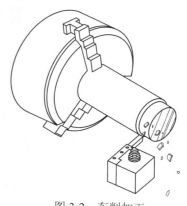

图 3-2　车削加工

钻削加工是用来生成圆孔。通常由具有双刀刃的旋转刀具完成。刀具沿着与其自身旋转轴平行的方向进给到工件中并形成圆孔（见图 3-3）。

在镗孔加工中，刀具是用来扩大已有的孔。它是用于产品制造最后阶段的精加工操作。

在铣削加工中，具有多刀刃的旋转刀具相对于材料慢慢移动，以生成水平面或垂直面。进给运动的方向与刀具旋转轴的方向垂直。旋转的铣刀完成速度运动（见图 3-4）。

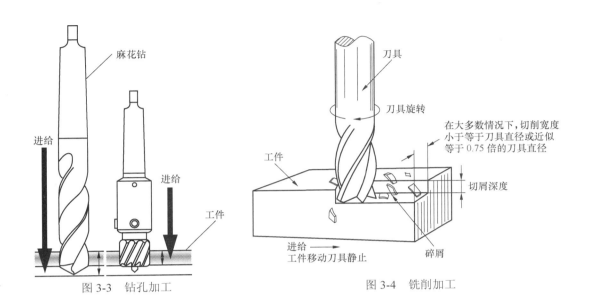

图 3-3　钻孔加工　　　　　　　图 3-4　铣削加工

加工技术概述

加工不是一个单一的过程，而是一系列过程。这些过程共同的特点是通过使用刀具形成从工件上切削下来的碎屑，称为金属切屑。为了完成加工操作，刀具和工件之间需产生相对运动。在大多数加工操作中，相对运动可通过产生切削速度的主运动和被称作进给的次运动来实现。刀具的形状、刀具对工件表面的切入，以及它们之间的运动组合在一起，生成了所期待的最终的工件表面（见图 3-5）。

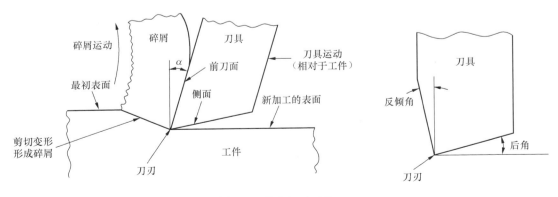

图 3-5　基本加工工艺

课文 B 车 削 机 床

车削机床通常被称为车床，有多种规格和外观设计。虽然大多数车床是水平式的车削机床，但是有时也使用垂直式的机床，垂直式机床用于加工大直径工件。车削机床也可根据所提供的控制形式进行分类。手动操作的车床要求操作者在车削操作中控制刀具的运动。车削机床也可以由计算机来控制，在这种情况下，它们被称为计算机数控（CNC）车床。计算机数控车床按照预先编制好的指令旋转工件并移动刀具，能进行高精度加工。在这类旋转机床中，旋转工件的主要部件和给进的刀具都是相同的。这些部件如下所示（见图 3-6）。

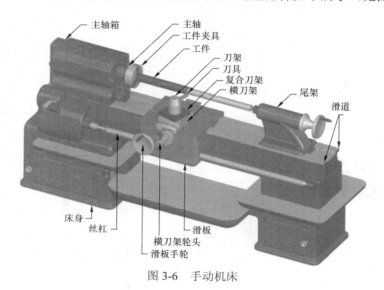

图 3-6 手动机床

床身

车削机床的床身只不过是一个大的基座，它位于地面或台面上并承载着机床的其他部件（见图 3-7）。

主轴箱

主轴箱位于机床的前部并固定在床身上。它包括为主轴提供动力的电机和驱动系统。主轴支撑并旋转工件，工件需通过工件固定装置或夹具，如卡盘或棘爪来固定（见图 3-8）。

图 3-7 床身

图 3-8 主轴箱

尾架

尾架位于机床的后部,并固定在床身上。当工件由主轴驱动时,尾架的作用是支撑工件的另一端并使之旋转。对于某些车削操作,工件无尾架支撑,以便可以将加工材料从尾端取下(见图3-9)。

滑板

滑板是工件旁滑动的平台,随着滑板的移动,它带动刀具切削材料。滑板位于床身的轨道上,被称为"滑道",通过电机驱动的丝杠或通过手轮向前运动(见图3-10)。

图 3-9 尾架

图 3-10 滑板

横刀架

横刀架固定在滑板的上部,带动刀具朝着或远离工件运动并改变切削深度。与滑板一样,横刀架也由电机或手轮驱动(见图3-11)。

复合刀架

复合刀架固定在横刀架的上部,被用来支撑刀具。刀具固定在复合刀架上的刀座中。复合刀架能够旋转并改变刀具相对于工件的角度(见图3-11)。

转塔

有些机床包括转塔,转塔上可安装多个刀具并将所需刀具旋转至切削工件的准确位置。转塔也可沿着工件移动并提供刀具切削材料。当大多数刀具在转塔中处于静止时,可使用活动工具加工。活动工具指的就是电动工具,如铣刀、钻头、铰刀和螺丝攻,它们可旋转并切削工件(见图3-12)。

图 3-11 横刀架轮头

图 3-12 转塔

第4单元 液压机械和锻压设备

课文A 液压机械

液压机械是利用流体动力来做功的机床和工具。在这类机床中，被称为压力流体的高压液体通过机床管路被传输到各种液压马达和液压缸内。

液压机械通过应用液压技术来运行。在液压机械中，液体是动力介质。另一方面，气动技术的基础是把气体作为动力传输、生成和控制的介质。

力和力矩的倍增

液压系统的一个基本特征是，在不需要机械齿轮组或杠杆系的情况下，通过改变两个相连汽缸间的有效面积，或改变泵和电机之间的有效位移，很容易地使力或力矩倍增（见图 4-1 和图 4-2）。这些例子通常是指涉及某种液压"传动比"的液压传动或液压静力传动。

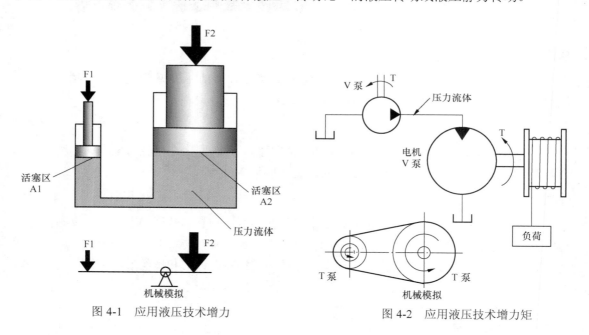

图 4-1 应用液压技术增力　　图 4-2 应用液压技术增力矩

液压回路

中位开放式回路使用可提供连续流量的泵。流体通过控制阀的开放式中位返回到油箱；也就是说，当控制阀处于中位时，它提供了一条返回油箱的通道，流体就不会被泵到高压（见图 4-3）。

中位封闭式回路给控制阀始终提供全压，无论阀门是否启动。泵可改变流动速率，在操作者开启阀门之前泵出很少的压力油。因此，阀门阀芯不需要开放式中位以便提供返回油箱的通路。多个阀门可以并联，使得各个阀门的系统压力相同（见图 4-4）。

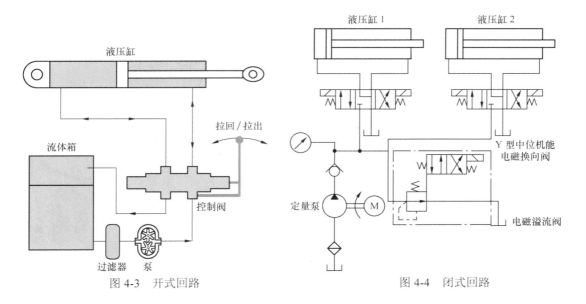

图 4-3 开式回路　　　　　图 4-4 闭式回路

开式回路和闭式回路

开环：泵的吸油口和马达的出油口（中间经由换向阀）与油箱连接。环这个术语与反馈有关；更准确的术语是开式"回路"和与之相反的闭式"回路"（见图 4-5）。

闭环：马达出油口直接连到泵的吸油口。为了保持低压侧的压力，回路中有个补油泵，用于给低压侧补充经过冷却和过滤的压力油。闭环回路通常用于移动应用中的液压静力传动（见图 4-6）。

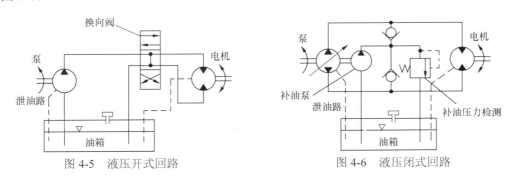

图 4-5 液压开式回路　　　　　图 4-6 液压闭式回路

优势：无需换向阀，灵敏度高，回路能在高压下工作。泵的旋转角可控制正负流向。

劣势：这种泵不容易实现其他液压功能。由于油量交换有限，冷却成了一个问题。

液压泵

液压泵给系统中的元件供应压力油。系统中的压力变化反应负载的大小。因此，压力为 5 000 磅/平方英寸的泵所产生的压力油可持续负担起 5 000 磅/平方英寸的负载。

泵具有比电机大 10 倍左右的功率密度（按体积计算）。它们由电动机或发动机驱动，通过齿轮、传送带或弹性橡胶联轴器相连以减少振动（见图 4-7）。

控制阀

换向阀通常设计成叠加式的，一个换向阀控制一个液压缸，从一个入口输入的流体可供应组套中所有阀门（见图 4-8）。

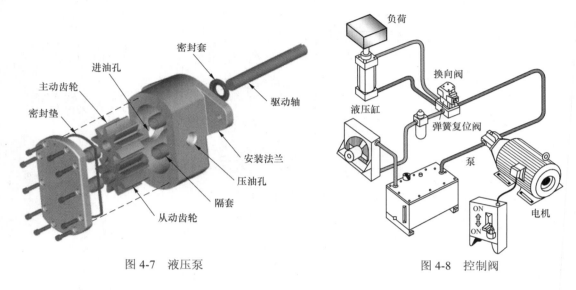

图 4-7　液压泵　　　　　　　图 4-8　控制阀

换向阀通过管路将流体送到期望的执行元件中。它们通常包括阀芯和铸铁或铸钢壳体。阀芯在壳内滑向不同位置，基于这个位置，阀芯环槽与壳体通道相交从而导通压力油。

课文 B　锻 压 设 备

锻造这个术语是指用定位挤压力使金属成形。冷锻造是在室温或接近室温的环境下进行的。热锻造是在高温下进行的，高温使金属更易于成形，且不易断裂。温锻造是在室温和高温锻造温度之间的温度下进行的。锻造零件的重量可从不到 1 千克到 170 公吨。

热锻造

热锻造的定义是在超过金属重结晶温度情况下对金属进行加工。热锻造的主要优点是当金属变形时，重结晶过程会消除应变硬化的影响（见图 4-9）。

温锻造

温锻造在节省成本方面有很多优点，这使得它成为一种广泛使用的生产方法。钢的温锻造温度从 800～1 800 华氏度不等。然而，1 000～1 330 华氏度之间的温锻造可能具有很

图 4-9　热锻造

大商业潜力。与冷锻造相比，温锻造具有以下潜在的优点：减少模具负荷、减少冲床负荷、增加钢的延展性、消除锻造前所需的退火过程、良好的可减少热处理的锻造属性（见图 4-10）。

冷锻造

冷锻造的定义是在低于金属重结晶温度的环境中进行加工，但是通常是在室温左右。如果温度超过熔化温度的 0.3 倍（以绝对温标计算），只能适应温锻造（见图 4-11）。

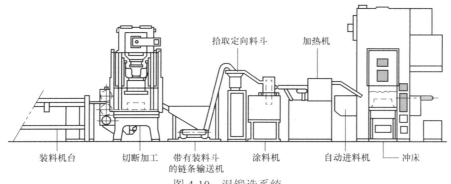

图 4-10 温锻造系统

图 4-11 冷锻造

过程

锻造是金属成形过程，用于生产大量相同零件及提高锻造金属的机械特性。当需要侧分型或型芯截面时，锻压件的设计会受到限制。所有型腔的绕口必须相当直且要足够大，以便能够取出锻模。锻造产品可小可大，可由钢（车轴）、黄铜（水阀）、钨（火箭喷管）、铝（飞机构件）或其他金属构成。一般的锻造过程包括：辊锻、型锻、钝齿啮合、开式模锻、飞边模锻、压锻、自动热锻和镦粗加工。

锻造的类型

锻造可以分为 3 种主要的方法：锤锻、压锻和模锻。

1. 锤锻：制造单个锻件的首选方法。在相对较小的区域里瞬间施压使某种金属或其他材料成形。锤子或撞锤通过间断锤打，给需要锻造的部分施加压力。锤子通常通过蒸汽压力或气压举高，然后从最高处下落。锤锻可产生各种各样的形状和大小，如果锤锻速度足够慢的话，可同时使晶粒高度细化。这个过程的缺点是经常需要精加工，因为难以获得精密的尺寸公差（见图 4-12）。

2. 压锻：这一过程与捏制相似，给要锻造的区域施加缓慢的连续压力。压力可触及材料深处，并通过冷锻或热锻来完成。冷压锻用于加工薄的退火材料，而热压锻用于加工大工件，如装甲钢板、火车头和重型机械。压锻比锤锻更经济（除了生产少量产品外），可以获得更精密的尺寸公差。压锻过程中的大部分压力可传递到工件上，这与锤锻加工不同，锤锻中的压力可被机床和基座吸收（见图 4-13）。

图 4-12　锤锻　　　　　　　　　图 4-13　压锻

3. 模锻：锻造可进行开式模锻和闭式模锻。在开式模锻中，模具既平又圆。大型锻件可通过对材料的不同部分连续地施压成形。液压机和锻造机都用于闭式模锻。在闭式模锻中，金属放入型腔凹处，可加工出上模和底模。当模具紧压到一起时，迫使材料填入型腔。飞边或多余的金属从模具之间挤出。闭式模锻与开式模锻相比，能生产出形状更加复杂的零件。就满足公差而言，模锻是最好的方法，可完全填满材料，并加工具有极少飞边的成品件。最终形状及冶金特性的改善取决于操作者的技能（见图 4-14）。

 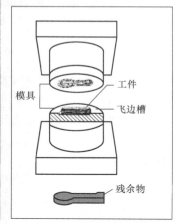

图 4-14　模锻

第 5 单元　计算机数控机床和计算机辅助制造设计简介

课文 A　计算机数控的基础

计算机数控

　　CNC 代表计算机数字控制，大约产生于 20 世纪 70 年代初期。在此之前，被叫作 NC，即数字控制。尽管大多数行业的人从未听说过这个术语，计算机数控已经以一种或更多种方式影响了各种生产过程。

CNC 如何工作

计算机数控机床由数控编程员来编程，它用加工文件来确定数控机床中每种切削工具移动的 X、Y 和 Z 坐标。这样就能对装入的零件进行切削、钻削、攻丝、钻孔、锪孔及倒角等操作。所有要求操作者在常规机床上进行的操作，都可以在数控机床上进行编程。一旦安装并运行，计算机数控机床便很容易保持运行。事实上，在长期的生产运行中，因为没什么可做的，所以计算机数控机床的操作者常常会感到无聊。使用某些计算机数控机床时，零件的装入过程都可实现自动化（见图 5-1）。

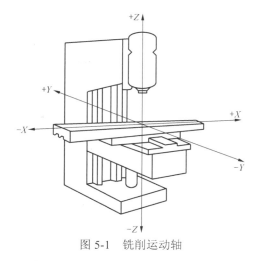

图 5-1 铣削运动轴

运动控制

所有类型的计算机数控机床都有一个共同之处，就是它们都有两个或更多的可编程的运动方向，称之为轴。运动轴或是线性的（沿着直线）或是旋转的（沿着圆形轨迹）。表明一台计算机数控机床复杂性的首要技术参数之一就是它的轴数。一般来说，轴越多，机床越复杂。

起初，以增量或步进方式旋转的步进电机作为运动控制技术为标准。位置精度决定了系统控制 X、Y 和 Z 轴实际位置精度。目前普遍使用三轴系统，但有的机床能控制 5～7 个运动轴（见图 5-2）。

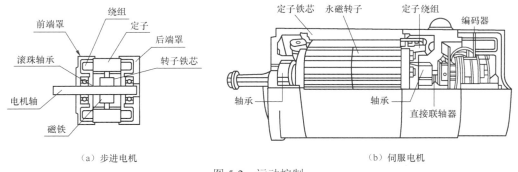

（a）步进电机　　　　　　　　　　（b）伺服电机

图 5-2 运动控制

可编程的附件

如果一台计算机数控机床只能在两个或多个轴的方向移动工件，那么它的作用不大。几乎所有的计算机数控机床都可进行几种方式编程。特殊类型的计算机数控机床通过适当的可编程附件能做许多事情。这里举出了某类机床的例子。

自动换刀装置

大多数加工中心在刀库中有许多刀具。当需要时，所需刀具能自动地放置在用于加工的主轴上（见图 5-3）。

主轴速度和启动

主轴速度（以每分钟转数计）很容易被限定，主轴能正向或反向接通旋转，当然也能关闭。

冷却剂

许多加工操作需要冷却剂来起到润滑及冷却作用。在加工周期内可以开始或停止提供冷却剂的动作（见图 5-4）。

图 5-3　自动换刀装置

图 5-4　冷却剂

计算机数控机床的程序

计算机数控机床的程序只不过是一组指令。它用句子一样的格式写入，控制器会一步一步地按顺序执行指令。

一系列特殊的计算机数控机床的指令字被用来传达机床要做什么。计算机数控机床的指令字以字母开始（如 F 代表进给速率，S 代表主轴速度，X、Y 和 Z 代表轴运动）。当以逻辑方式排列到一起时，一组计算机数控机床的指令字便组成了类似句子的指令（见图 5-5）。

对于已知的计算机数控机床，只有 40～50 个指令字经常使用。所以如果您将学习编写计算机数控机床程序比作是学习一门只有 50 个单词的外语的话，那么计算机数控机床编程似乎并不十分难学。

计算机数控机床的控制器

计算机数控机床的控制器会解译数控程序并按顺序使一系列指令生效。当读程序时，控制器会使适当的机床功能生效并产生轴的运动，总之，会执行程序中的指令。一般来说，计算机数控机床的控制器可允许操作机床的全部功能（见图 5-6）。

图 5-5　数控编程

图 5-6　计算机数控控制

课文 B 用 Alibre 计算机辅助制造软件设计零件

Alibre Design 软件使得运用 Alibre 计算机辅助制造软件来加工零件变得简单。以麦克软件技术为基础，Alibre 计算机辅助制造软件在 Alibre 的龙头产品 Alibre Design 专业版 10.0 中。Alibre 计算机辅助制造软件拥有大量的预设及可定制的软件工具，还可选某些输出软件。Alibre 计算机辅助制造软件可支持 2.5 轴、3 轴和钻削操作。Alibre 计算机辅助制造软件是实现从概念设计到成品制造的极佳工具。

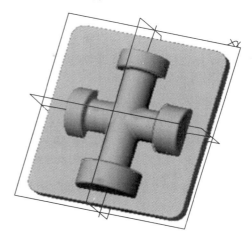

创建一个零件

第一步：在加工中使用 Alibre 计算机辅助制造软件进行加工的第一步是创建一个零件。

图 5-7 以设计零件开始

例如，我们将使用 Alibre 计算机辅助制造软件的零件安装程序中的一个部件作为示例。进入 C:\Program Files\Alibre CAM 1.0\Tutorials，找到名为 3AxisExample1.AD_PRT 的零件程序。作为选择，你便可以创建自己的所用零件（见图 5-7）。

第二步：打开 Alibre Design 中的零件程序后，继续进入到屏幕顶端 Alibre 计算机辅助制造软件浏览器的菜单栏。选择 Alibre 计算机辅助制造软件的浏览器来激活 Alibre 计算机辅助制造软件的界面（见图 5-8）。一旦进入 Alibre 计算机辅助制造软件，单击该软件资源管理器中的"建立（Setup）"标签。单击"建立加工（Set Up Machine）"按钮来改变刀具的最初位置（见图 5-9）。

图 5-8 选择 Alibre 计算机辅助制造软件

图 5-9 选择"建立（Setup）"标签

第三步："后处理程序选项设置（Set Post-Processor Options）"菜单可让你设置一组选项，如果愿意的话，还可让你选择查看机器代码的程序（见图 5-10）。

第四步：下一步是创建你的工料或装载先前已创建的工料。"创建/装载工料"按钮以及"工料中设置零件"按钮允许你设置最初工料的尺寸，并在工料中设置零件（见图 5-11）。

图 5-10 选择"后处理程序选项"设置菜单

图 5-11 选择创建/装载工料按钮

创建工具

现在来创建工具。单击"刀具"标签查看新的选项（见图 5-12）。

第五步：单击"创建/选择刀具"选项。在"刀具"选项的对话框顶部，你可以从 4 种标准的刀具中做出选择（见图 5-13）。单击其中的一个工具，然后设置刀具的属性和尺寸。完成创建刀具后，单击"保存新刀具"选项，然后单击"确定"选项，把新刀具加入表中。提醒一下，你也能装载刀库或创建你自己的刀库，以便把重复性工作减到最少。

图 5-12　选择刀具标签

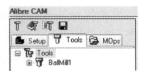
图 5-13　创建/选择刀具按钮

你可以通过"创建/选择刀具"对话框调整你的工具（见图 5-14）。

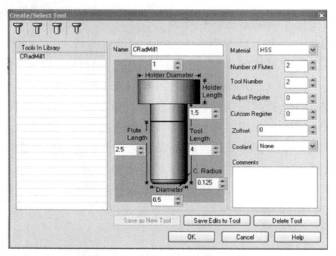
图 5-14　创建/选择刀具对话框

增添加工操作和模拟刀具路径

在进行完零件、工料和工具选项的设置后，该到设置我们想要执行的加工类型了。在这个例子中，我们选择平行精加工作为加工类型。

第六步：选择合适的刀具后，单击 Mops（加工操作的缩写）标签。找到铣削方法（Milling Methods）按钮（见图 5-15）。单击此按钮时，会出现带有"水平粗加工（Horizontal Roughing）、平行精加工（Parallel Finishing）和设置文件（Profiling）"选项的下拉菜单。

第七步：通常先用水平粗加工生成零件的大概形状。平行精加工通常用于水平粗加工之后，通过除去粗加工产生的多余材料来增加形状的精确度。单击"平行精加工"项，平行精加工对话框就会弹出（见图 5-16）。根据应用情况设置合适的选项，然后单击"生成（Generate）"按钮。

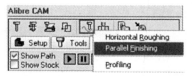
图 5-15　查找"铣削方法"按钮

接下来，准备查看模拟的刀具路径。单击"播放（Play）"按钮，观察所创建的刀具路径

（见图 5-17）。

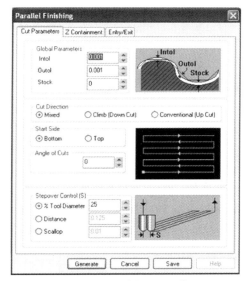

图 5-16　平行精加工对话框

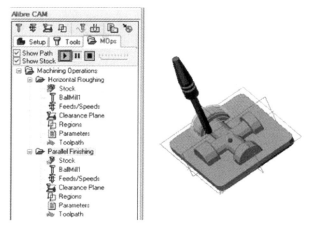

图 5-17　机术操作标签

后加工

在创建完刀具路径后，必须根据所要使用的机床，将其以适当的格式输出。Alibre 计算机辅助制造软件提供了大量后加工程序供你选择。

第八步：单击"后处理程序（Post Processor）"按钮（见图 5-18），选择后处理程序。设置输出目录，单击"确定（OK）"按钮，文件生成。

你可以用输出的文件进行零件加工了。

恭喜！你刚刚使用 Alibre 计算机辅助制造软件创建了第一个机器代码（见图 5-19）。通过监视一些选项，如进给/速度设置、间隙控制、Z 轴控制、接近值和啮合值，你可以微调设置。如果你需要更多的教程或综合的帮助，可以经常查看 Alibre 计算机辅助制造软件帮助功能来获得更多的信息和提示。

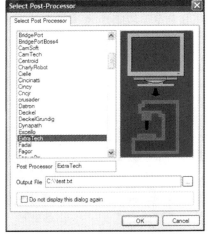

图 5-18　选择后处理程序

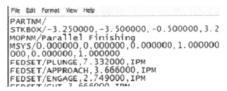

图 5-19　创建机器代码

第6单元 机械制图

课文A 工程图（Ⅰ）

技术制图，又称制图，是由建筑师、室内设计师、制图员、设计工程师以及相关专业人士用来创建标准技术图纸的学科。布局、线的粗细、文本长度、符号、视图投影、画法几何、标注尺寸和记号的标准和规定常用于创建图纸，这些图纸只能用一种方法来圆满地解读。

从事绘图的人被称为制图员。在某些领域，制图员被称为制图技术员、绘图员或绘图人。制图员绘制的技术图是一种专门的图形交流语言。（见图6-1）。技术图与一般绘图的不同之处在于它解读的方法不同。一般绘图可以有多个目的和意义，而技术图的目的是简单明了地表示出所需的全部参数，以便形成与实物相同的概念。（见图6-2）。

图6-1 制图员在工作

图6-2 复制工程图

技术图的种类

工程制图

工程制图是技术制图的一种，用工程原理创建，用于完整清晰地表达工程项目的要求。

工程图依据标准规定绘制，这些标准包括：布局、术语、图示、外观（如字型和线型）、尺寸等（见图6-3）。

工程制图的目的是准确且毫不含糊地捕捉产品或部件的全部几何特征。工程制图的最终目的是表达出让制造商能够生产该零件所需的全部信息。

剖视图

剖视图是一种技术图解，为使模型内部的特征显露出来，会有选择地除去三维立体模型的表面的元件，但不会完全破坏外部背景（见图6-4）。

剖视图的目的是使观察者能看到一个非透明实体的内部，并不是让内部物体透过外表面显现出来，而是把部分外部物体简单地除去。这就产生了好像有人把一个物体删除了一部分或把一个物体切成了几部分的视觉效果。剖视图避免空间排序的含糊不清，形成了前景与背景物体的鲜明对比，使得对空间排序的理解变得容易。

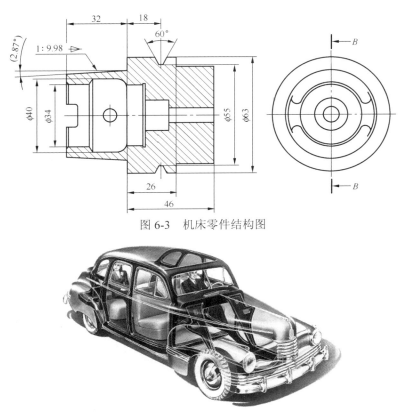

图 6-3　机床零件结构图

图 6-4　纳什 600 的剖视图

分解图

　　分解图是技术图的一种，它展示了物体各部分的装配关系或装配顺序。它展示的物体各部分是按一定距离略微分开的或以三维分解图的方式悬置于周围的空间。物体的表示形式是，好像物体内部发生了微小的、可控的爆裂，使得物体的各部分离开最初的位置，并以等距离分开（见图 6-5）。

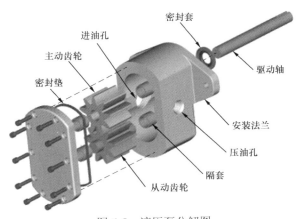

图 6-5　液压泵分解图

分解图可显示机械零件或其他零件的既定装配顺序。通常状况下，机械系统中最靠近中心的零件或是其他零件需要装配其上的主要零件首先装配。分解图还有助于表明零件的拆卸顺序，通常外部零件首先被拆卸。

专利图

专利图是体现专利发明的一种技术图，它能够体现发明的本质。专利图必须体现出专利申请中所宣称的发明的每一个特点，并根据专利局所要求的规范以特定的形式绘制（见图6-6）。

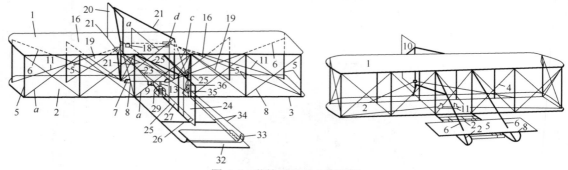

图 6-6　莱特兄弟的专利图纸

当需要通过图纸来理解发明的性质时，法律要求专利申请人提供发明图纸。因此图纸要与专利申请函一并提交。这适用于除物质组成或工艺方法之外的几乎全部发明。但对于很多工艺方法而言，专利图也很有用。

课文 B　工程图（Ⅱ）

技术图解

技术图解是用图解形象表达技术性信息。技术图解可以是部件工程图或图表。技术图解的目的是生成具有表达力的图像，有效地通过视觉方式向看图者传达某些信息（见图6-7）。

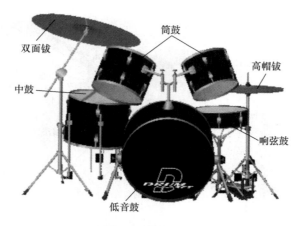

图 6-7　技术图解

技术图解的主要目的是向不太懂技术的看图者介绍或解释产品部件。可视图像在尺寸和比例方面力求精确，使看图者对物体是什么以及物体有什么功能有总体的印象，以提高他们的兴趣和理解。

技术草图

草图是一种快速绘制的徒手图，并不是用作成品图。一般情况下，草图是一种快速记录想法并用于以后使用的方法。建筑师的草图主要是一种在开始成品建设前尝试不同想法和组合的方法，尤其当加工成品件既昂贵又耗时时，更需在加工前着手绘制草图（见图6-8）。

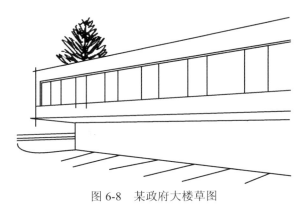

图 6-8　某政府大楼草图

例如，建筑素描作为一种绘图形式，类似于比喻修辞，是建筑师用来辅助协同设计的一种表达方式。

基本绘图纸的大小

如下面图纸系列所示，按次序后一型号绘图纸的高度是前一型号的二倍（见表6-1）。

表 6-1　　　　　　　　　基本绘图纸大小

图 纸 类 型	尺寸（宽×高）	图 纸 类 型	尺寸（宽×高）
A 号	8.5 英寸 × 11.0 英寸 22 厘米 × 28 厘米	B 号	11.0 英寸 × 17.0 英寸 28 厘米 × 43 厘米
C 号	17.0 英寸 × 22.0 英寸 43 厘米 × 56 厘米	D 号	22.0 英寸 × 34.0 英寸 56 厘米 × 86 厘米
E 号	34.0 英寸 × 44.0 英寸 86 厘米 × 112 厘米	F 号	44.0 英寸 × 68.0 英寸 112 厘米 × 173 厘米
G 号	68.0 英寸 × 88.0 英寸 173 厘米 × 224 厘米	H 号	88.0 英寸 × 136 英寸 224 厘米 × 345 厘米

手工制图

手工制图的基本步骤是把一张纸铺在有直角和直边的光滑表面上，通常是放在制图桌上（见图6-9）。可滑动的直尺即丁字尺一边固定，另一边可在制图桌和图纸表面上滑动。

图 6-9　制图台

通过移动丁字尺并沿着丁字尺边挪动铅笔或针笔可绘制出"平行线",但丁字尺最典型的用途是作为固定绘图仪器如三角板的工具。现代制图台(已很大程度上被计算机辅助设计工作站所取代)配备了平行规,平行规固定在制图台的两侧,可在大张图纸上滑动。

此外,制图员也用一些工具来绘制曲线和圆。其中最主要的是圆规,用来勾勒简单的弧线和圆,然后是曲线板,即具有复杂曲线形状的塑料板。曲线规是一块是表面包有橡胶的金属铰链,可人为地弯成各种曲线(见图6-10)。

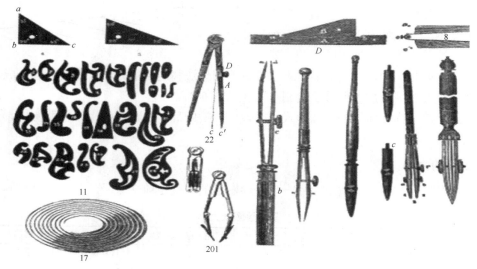

图6-10 技术图工具

计算机辅助设计

今天,制图技术已大大地实现了自动化,并且计算机辅助设计(CAD)系统使得制图速度得到了提高。计算机辅助设计系统是用计算机技术来辅助零件或产品,甚至是整个建筑的设计和绘图(见图6-11)。

图6-11 计算机辅助设计

可在二维或三维空间下制图。制图是工程图或机械图的统称,同时它也是工艺美术的分支学科。在用二维图表示三维复杂物体时,传统的表达方式是在合适的角度绘制出该物体的三个投影视图。

第 7 单元　电器元件与电路

课文 A　电器元件及符号

电子部件是基本的电子元件，通常是以离散形式封装，有两个或更多的连接引线或金属垫片。这些部件通常被焊接到印制电路板上，连接在一起形成一个具有特定功能的电子电路（例如放大器、无线电接收器或振荡器）。

有大量电子部件也就有同等数量的代表它们的符号。认识常见的部件并了解它们的功用是非常重要的。

二极管

二极管基本上就是单向电流阀。它允许电流单向流动（从正极到负极），而不能沿反方向流动。大多数二极管外表与电阻器相似，一端画有漆线表明电流流动方向（白边表明是负极）。如果负极边连到电路的负极端，电流就会流动。如果负极边连到电路的正极端，电流就不会流动（见图 7-1）。

晶体管

有两种类型的标准晶体管 NPN 和 PNP，它们有不同的电路符号。晶体管或许是近十年来最重要的发明。它执行两个基本功能：（1）作为开关接通或切断电流；（2）作为放大器，将输入信号放大产生输出信号（见图 7-2）。

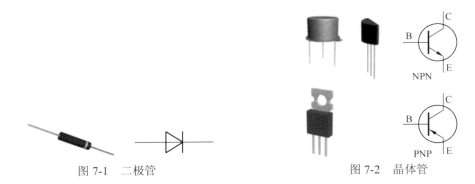

图 7-1　二极管　　　　图 7-2　晶体管

电阻器

像二极管和继电器一样，电阻器是另一个电子部件，他们应该在安装人员的部件柜中占有一席之地。它已成为移动电子装置中必须安装的，无论是用于门锁电路、定时电路、远程启动装置、发光二极管电路还是用于硬化电容器的放电（见图 7-3）。

电阻器"阻止"电流流动。电阻值越高（测量单位为欧姆），电流就越小。

电容器

在电路中用于存储电荷的装置。电容器的功能更像电池，但充放电速度比电池快（然而电池可以储存更多的电荷）（见图7-4）。

基本电容器由两个导体组成，导体由绝缘体或电介质分离。

晶闸管

晶闸管是一种固态半导体器件，由四层N型和P型材料交替组成。它们起到双稳态开关的作用，当门极收到电流脉冲时晶闸管开始导电，而且只要处于正向偏置（也就是说，只要整个晶闸管的电压没有处于反向），晶闸管就会继续导电（见图7-5）。

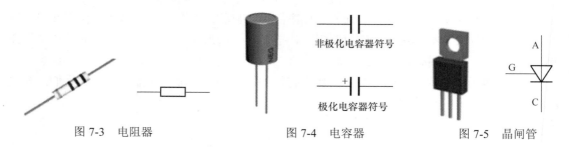

图7-3 电阻器　　　　　　图7-4 电容器　　　　　　图7-5 晶闸管

有些人把可控硅整流器和晶闸管定义为同义词。

发光二极管

发光二极管（LED）是一种电子光源。发光二极管是以半导体二极管为基础。当二极管正向偏置（接通）时，电子在空穴中重新组合，能量以光的形式释放，此效果被称为电致发光。光的颜色由半导体的能隙决定（见图7-6）。

发光二极管和传统光源相比有许多优点，如能耗较低、寿命较长、强度改进、尺寸较小、开关较快。

继电器

继电器是一个电动开关。流过继电器线圈的电流会产生一个磁场来吸引控制杆并改变开关触点的吸合。线圈中允许或阻止电流流过，所以继电器可处于开关两个位置，作为双向开关（转换）开关（见图7-7）。

图7-6 发光二极管　　　　　　图7-7 继电器

课文B　电路的基本概念

此图以框图形式表明了电路的基本形式（见图7-8）。它由电源、消耗能量的某类负载以及连接电源和负载的电导体组成。

电源有两个接线端子，被指定为正极（+）和负极（-）。如图7-8所示，只要电源到负载以及负载到电源的电路连接不断开，电子就会从电源负端流出，经过负载，然后流回电源的正端。箭头表明了电子流流经电路的方向。由于电子总是朝着同一方向流经电路，所以它们的运动被称为直流。

电源可以是任何电能来源。在实际中，总体上有3种电能来源：它可以是电池、电力发电机或某类电子电源。

负载是被供电的装置或电路。它可以像灯泡一样简单或像现代化高速计算机一样复杂。

由电源所提供的电量有两个基本特征，分别称为电压和电流。它们被定义如下。

电压

电"压"使得自由电子流经电路，也被称为电动势（EMF）。

有两种主要类型的电源，即电压源和电流源。电源既可以是独立的也可以是由其他量决定的。独立的电压源可保持电压大小而不受任何其他量的影响。同样，独立的电流源可保持电流大小而不受任何其他量的影响（见图7-9）。

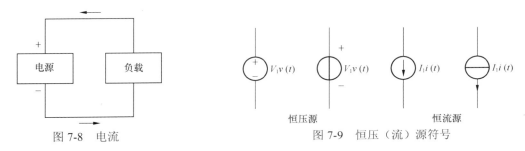

图7-8 电流　　　　　　　　　图7-9 恒压（流）源符号

有些电压（电流）源的电压（电流）值会随着一些其他变量的变化而变化。它们被称为非独立电压（电流）源或可控的电压（电流）源（见图7-10）。

电流

在大量的如电阻器和电源的装置中，终端通过连接导线连接起来，这被称为电路。这些导线汇聚于节点处，这些装置所在的电路被称为支路（见图7-11）。

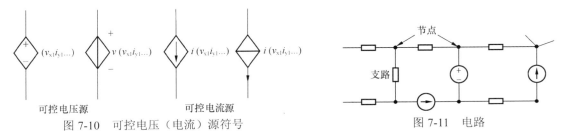

图7-10 可控电压（电流）源符号　　　　图7-11 电路

电路的普遍问题是在明确了电源的大小后，需确定全部支路上电流和电压的大小。这样的问题通常被称为电路分析。

电阻

当把电压施加于一段金属丝时（见图7-12），流经金属丝的电流 I 与金属丝两端的电压 V 成正比。这个属性被称为欧姆定律，可写为：$V = IR$ 或 $V = I/G$。在这里，R 被称为电阻，G 被称为电导率。同一段金属丝的电阻 R 和电导率 G 的关系是 $R = 1/G$。电阻的单位以欧姆（Ω），

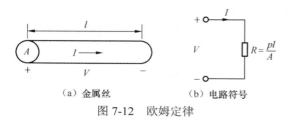

(a) 金属丝　　(b) 电路符号

图 7-12　欧姆定律

电导率的单位是西门子（S 或Ω）。

电路中电压、电流和电阻的关系对于任何电路或装置的运行而言都是很基本的。可以说，流经电路的电流量与所施加的电压成正比，而与电阻成反比。明确的定义是 1 伏特的电压可以推动 1 安培的电流流经 1 欧姆的电阻。

第 8 单元　单片机

课文 A　单片机及其电路简介

单片机是指一个集成在一块芯片上的完整计算机系统。即使它的大部分功能集成在一块小芯片上，但是它具有一个完整计算机所需要的大部分部件：中央处理器（CPU）、内存、内部和外部总线系统，大部分还会具有外存。同时它还集成诸如通信接口、定时器、实时时钟等外围设备。而现在最强大的单片机系统甚至可以将声音、图像、网络和输入输出复杂系统集成在一块芯片上（见图 8-1）。

单片机是单片集成电路，普遍具有如下特征（见图 8-2）。

图 8-1　芯片

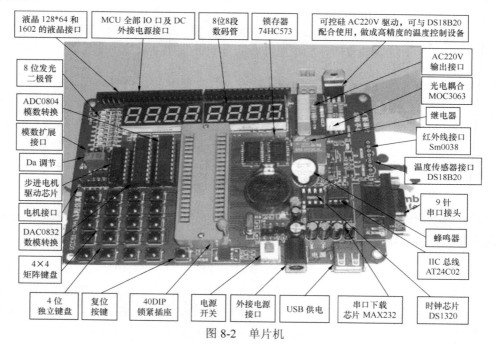

图 8-2　单片机

中央处理器——从小而简单的 4 位处理器到复杂的 32 位或 64 位处理器；

输入/输出接口，如串行口（异步串行接口）；

其他串行通信接口，如串行外围接口和用于系统连接的控制器区域网络；

外设，如定时器和看门狗；

随机存储器，用于数据存储；

只读存储器、可擦可编程只读存储器、电可擦可编程只读存储器和闪存，用于程序存储；

时钟发生器——经常采用石英晶体定时振荡器、谐振器或 RC 电路；

模数转换器。

供应商可提供各种各样的单片机结构。8051、Z80 和 ARM 系列产品是其中的主要衍生品。多数单片机使用哈佛结构：用于指令和数据的存储总线分离，可使存储同时进行。与通用中央处理器相比，单片机没有地址总线和数据总线，因为它同中央处理器一样可将随机存储器和永久性存储器集成在一块芯片上。由于管脚很少，芯片的体积就会很小，封装起来更加便宜（见图 8-3）。

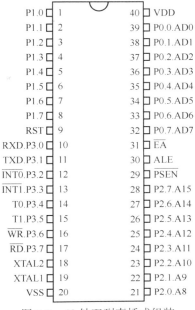

图 8-3　40 针双列直插式组装

单片机也被称为微控制器，这是因为它最早用于工业控制领域。单片机是由芯片内仅有 CPU 的专用处理器发展而来的。最早的设计理念是通过将大量外围设备和 CPU 集成在一个芯片中，使计算机系统更小，更容易集成到复杂的且对体积要求严格的控制设备当中。INTEL 的 Z80 是最早按照这种思想设计出的处理器，从那时起，单片机和专用处理器的发展便分道扬镳。

单片机比专用处理器更适合应用于嵌入式系统，因此它的应用更加广泛。与应用在个人计算机中的通用型微处理器相比，它强调自供应（不用外接硬件）和节约成本。现代人类生活中所用的几乎每件电子和机械产品中都有单片机。手机、电话、计算器、家用电器、电子玩具、掌上电脑以及鼠标等电脑配件中都配有 1~2 台单片机。而个人电脑中也有为数不少的单片机在起作用。汽车一般配备 40 多台单片机，复杂的工业控制系统甚至可能有数百台单片机在同时起作用。单片机的数量不仅远超 PC 机和其他计算机装置的数量，甚至比人类的数量还要多。

课文 B　MCS51 系列单片机简介

早期的单片机都是 8 位或 4 位的。其中最成功的是 INTEL 的 8031，因为简单可靠而性能不错获得了很大的好评。此后在 8031 上发展出了 MCS51 系列单片机系统。基于这一系统的单片机系统直到现在还在广泛使用。随着工业控制领域要求的提高，开始出现了 16 位单片机，但因为性价比不理想并未得到很广泛的应用。20 世纪 90 年代后随着消费电子产品大发展，单片机技术得到了巨大的提高。随着 INTEL i960 系列特别是后来的 ARM 系列的广泛应用，32 位单片机迅速取代 16 位单片机的高端地位，并且进入主流市场。而传统的 8 位单片机的性能也得到了飞速提高，处理能力比起 80 年代提高了数百倍。目前，高端的 32 位单片机频率已经超过 300MHz，性能直追 90 年代中期的专用处理器，而普通的型号出厂价格跌落至 1 美元，最高端的型号也只有 10 美元。当代单片机系统已经不再只在裸机环境下开发和使用，大量专用的嵌入式操作系统被广泛应用在全系列的单片机上（见图 8-4）。

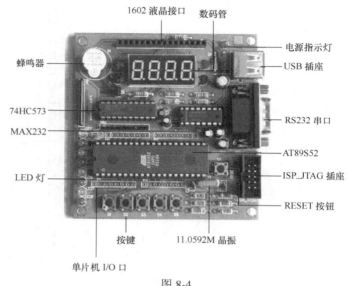

图 8-4

现在举例说明 MC51 系列单片机，AT89C2051 单片机是 51 系列单片机的一个成员，是 8051 单片机的简化版。内部自带一个 2k 字节可编程 EPROM 的高性能微控制器。它与工业标准 MCS-51 的指令和引脚兼容，因而是一种功能强大的微控制器，它对很多嵌入式控制应用提供了一个高度灵活有效的解决方案。AT89C2051 有以下特点：2k 字节 EPROM、128 字节 RAM、15 根 I/O 线、2 个 16 位定时器/计数器、5 个向量二级中断结构、1 个全双向的串行口并且内含精密模拟比较器和片内振荡器，具有 4.25V 至 5.5V 的电压工作范围和 12MHz/24MHz 的工作频率，同时还具有加密阵列的二级程序存储器加锁、掉电和时钟电路等。此外，AT89C2051 还支持两种软件可选的电源节电方式。空闲时，CPU 停止，而让 RAM、定时/计数器、串行口和中断系统继续工作。可掉电保存 RAM 的内容，但可使振荡器停振并禁止所有的其他功能直到下一次硬件复位。

单片机的时钟信号用来提供单片机片内各种微处理器操作的时间基准，时钟信号通常用两种电路形式得到：内部振荡和外部振荡。MCS-51 单片机内部有一个用于构成振荡器的高增益反向放大器，引脚 XTAL1 和 XTAL2 分别是此放大电器的输入端和输出端，由于采用内部方式时，电路简单，所得的时钟信号比较稳定，实际使用中常采用这种方式，如图 8-5 所示，在其外接晶体振荡器（简称晶振）或陶瓷谐振器中构成了内部振荡方式，片内高增益反向放大器与作为反馈元件的片外石英晶体或陶瓷谐振器一起构成了一个自激振荡器并产生振荡时钟脉冲。图 8-5 中外接晶体以及电容 C1 和 C2 构成了并联谐振电路，它们起稳定振荡频率、快速起振的作用，其值均为 30pF 左右，晶振频率选 12MHz。

为了初始化单片机内部的某些特殊功能寄存器，必须采用复位的方式，复位后可使 CPU 及系统各部件处于确定的初始状态，并从初始状态开始正常工作。单片机的复位是靠外电路来实现的，在正常运行情况下，只要 RST 引脚上出现两个机器循环时间以上的高电平，即可引起系统复位，但如果 RST 引脚上持续为高电平，单片机就处于循环复位状态。复位后系统将输入/输出（I/O）端口寄存器置为 FFH，堆栈指针 SP 置为 07H，SBUF 内置为不定值，其余的寄存器全部清 0，内部 RAM 的状态不受复位的影响，在系统上电时 RAM 的内容是不定的。复位操作有两种情况，即上电复位和手动（开关）复位。本系统采用上电复位方式。如图 8-5 所示（见原文图），R0 和 C0 组成上电复位电路，其值 R 取为 4.7kΩ，C 取为 10μF。

| 第 9 单元　电机介绍 |

课文 A　电机基础（I）

电机的功能是把电能转换成机械能。因此，电机也分类为机电装置。

电磁力的基本概念

电磁力

永久磁铁产生的磁通方向总是 N 极指向 S 极。

当导体置于磁场，且有电流通过时，磁场和电流相互作用产生了力。这个力被称为"电磁力"（见图 9-1）。

弗莱明左手规则决定了电流、磁力和磁通的方向。像图 9-2 所示的那样，伸出左手的拇指、食指和中指。

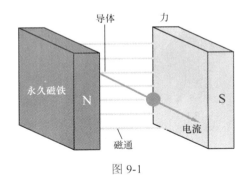

图 9-1

图 9-2

当中指指向电流方向，食指指向磁通方向时，拇指指向力的方向。

由电流产生的磁场

电流产生的磁场和永久磁体相互作用产生电磁力。

当导体中的电流朝着注视者的方向流过时，根据右手螺旋定则，电流周围会产生逆时针方向旋转的磁场（见图 9-3）。

磁力线的干涉

电流产生的磁场和永久磁体相互干涉（见图 9-4）。

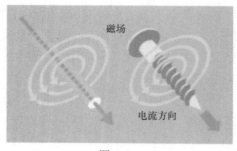 

图 9-3　　　　　　　　　　　图 9-4

同一方向分布的磁力线会增加力,而相反方向分布的磁通会减小力。

电磁力的产生

磁力线像松紧带一样具有通过自身张力回到直线状态的特性。

因此,导体会受到磁场力的作用从磁力强的位置移动到磁力弱的位置（见图 9-5）。

扭矩的产生

电磁力可通过方程求得:

F（力）= B（磁通密度）· I（电流）· L（导体长度）

图 9-6 展示了单圈导体置于磁场获得的扭矩。

单圈导体所获得的扭矩可通过方程求得:

$T' = F \cdot R$

- T'（扭矩）
- F（力）
- R（离导体中心的距离）

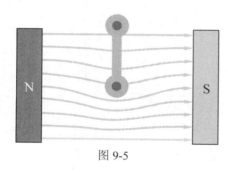 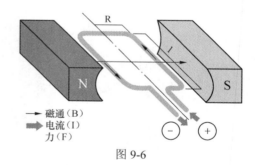

图 9-5　　　　　　　　　　　图 9-6

下面是两个导体的扭矩大小方程:

$T = 2T' = 2 \cdot F \cdot R$

电机的运动原理

1. 置入永久磁铁

放置永久磁铁,不同极性相对,这样就产生了平行磁场（见图 9-7）。

2. 置入电枢

在磁场的平行方向置入电枢,同时产生了磁化现象,①是 N 极、②和③是 S 极。

然后,由于受到磁场的吸引和排斥,电枢开始顺时针旋转（见图 9-8）。

图 9-7

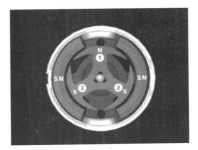

图 9-8

3．电枢磁化

加上绕阻，接通电流，电枢会被磁化（见图 9-9）。

4．保持旋转

为了保持电枢旋转，应改变电流方向，使 N 极总在电枢的上部，S 极总在电枢的下部（见图 9-10）。

图 9-9

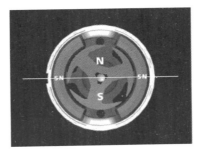

图 9-10

5．切换电流方向

为了切换电流方向，应把线圈连接到三个独立的柱形金属条（换向器）上（见图 9-11）。

6．置入电刷

为了保持顺时针旋转，需在左侧电刷加正（+）电压，右侧电刷加负（−）电压。

如果需要逆时针旋转，需在左侧电刷加负（−）电压，右侧加正（+）电压，或使磁铁极性相反（见图 9-12）。

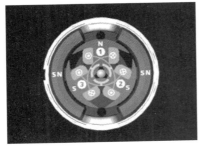

图 9-11

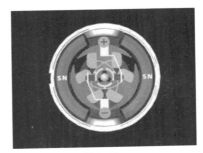

图 9-12

课文 B 电机基础（Ⅱ）

电机主要部件的功能

1. 产生磁场

永久磁铁保持磁力并固定在电机壳上。电机壳本身产生磁路（磁场），并以铁芯和永久磁铁作为磁力流动的路径（见图 9-13）。

2. 流经电流

电流从电机的一个接线端（向前伸出的部分）流入。

电流通过电刷和换向器流入绕组，再通过电刷和换向器从电机的另一个接线端流出（见图 9-14）。

3. 旋转输出

电流流入永久磁铁产生的磁场，同时产生电磁力。换向器在恰当的时机切换电流方向以保持电机连续旋转。通过传动轴产生输出（见图 9-15）。

图 9-13

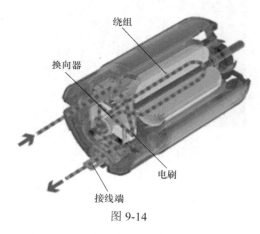

图 9-14

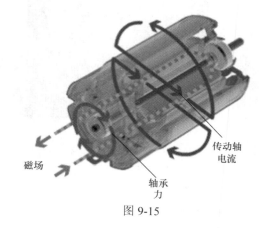

图 9-15

电机部件说明（见图 9-16～图 9-20）

图 9-16 整个电机

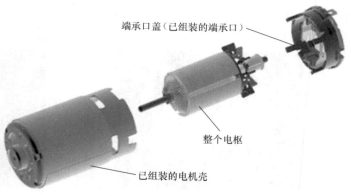

图 9-17 整个电机的分解图

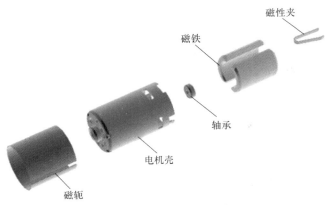

图 9-18　电机壳分解图

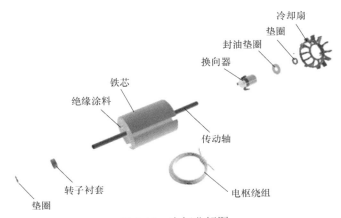

图 9-19　电枢分解图

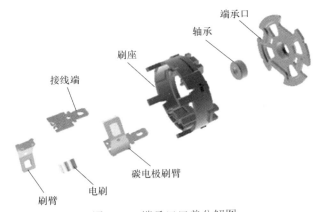

图 9-20　端承口口盖分解图

电机性能术语表

术　语	符　号	单　位
输入	P	瓦
输出	P	瓦
最大输出	P max.	瓦
电压	V	伏特
电流	I	安培

续表

术 语	符 号	单 位
空载电流	I0	安培
失速电流	Is	安培
效率	η	百分数
最大效率	η max.	百分数
转速	N	转/分
空载转速	N0	转/分
扭矩	T	牛顿·米，克·厘米
失速扭矩	Ts	牛顿·米，克·厘米

电机的总体使用说明

在使用电机时，请务必阅读以下一般性说明以确保电机使用的安全、可靠和正确。

警告

1. 不要把导线或电机接线端插入家用插座中，否则可能触电。
2. 当通入电流时，不要触摸带电部件，如载流接线端，否则可能触电。
3. 当通入电流时，手或手指远离旋转部件，否则可能受伤。
4. 当通电时，不要锁紧电机的传动轴，因为即使锁紧时间很短也会造成过多的热量积聚，从而导致电机烧毁。
5. 根据电机的运行状态（安装状态、负载和环境温度），电机可能会产生过多的热量。切勿因此烧伤自己。

第 10 单元　可编程逻辑控制器简介

课文 A　可编程逻辑控制器

这部分仅仅展示了可编程序逻辑控制器的一小部分性能。像计算机一样，PLC 可执行定时功能（相当于时滞继电器）、鼓测序以及其他的高级功能，精度和可靠性比使用机电逻辑器件要好得多。大多数 PLC 有超过六输入和六输出的容量。图 10-1 展示了单个艾伦-布拉德利可编程序控制器的几个输入和输出模块。

PLC 中的每个模块都有 16 个"点"的输入或输出，具有监视和控制数十台装置的能力。PLC 可安装到控制柜中，只需占用很小的空间，特别是考虑到执行相同功能的机电式继电器所需要的空间（见图 10-2）。

PLC 的一个机电式继电器所不具备的优点是可通过数字化计算机网络进行远程监视和控制。可编程序控制器（PLC）只不过是一种专用的数字化计算机，因此它能够很容易地与其他计算机通信。图 10-3 是一台个人计算机，显示了由 PLC 控制的真实液面控制过程图像（是用于城市污水处理系统的一个泵站）。实际的泵站距离个人计算机显示的画面有数英里之远。

可编程逻辑控制器的结构如图 10-4 所示。结构包括可输入信息的各种输入开关装置，如按钮、限位开关和继电器触点。还有输出装置可连接的终端，如螺线管、继电器线圈及显示灯。输入与输出之间没有直接连线。相反，输入的开关状态被转换为逻辑电平信号，作为控制器中数字化计算机的输入信号。根据当前和过去的输入信号状态计算机中存储的程序会决定应该激活哪些输出。来自于控制器的逻辑电平输出信号被转换为电压电平，用于激活或关闭各种输出装置。

图 10-1　可编程逻辑控制器

图 10-2　控制柜

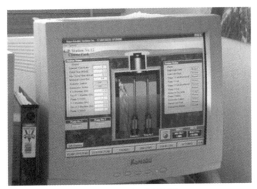

图 10-3　运程监视

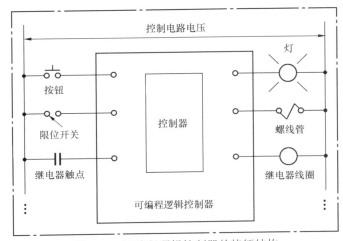

图 10-4　可编程逻辑控制器的特征结构

可编程逻辑控制器的软件由操作系统、基本软件和用户程序组成。操作系统包含用于内部系统功能的所有语句和陈述。基本软件有一个运行基本功能的灵活接口。基本软件也包含一些功能块。用户程序是由用户编制的所有语句和陈述的总和。用户程序由不同类型的模块组成，如组织模块、程序模块、功能模块、顺序模块和数据模块。它们用于组成用户程序，每1个模块用于完成不同的任务。

许多控制应用要求各种输出转换（开或关）作为大量输入装置的状态（开或关）功能。这种类型的控制被看作是逻辑控制的转换控制。在自动机床和加工过程中很有吸引力，因为这些机床和加工过程要遵循一套操作程序。举一个传送线的例子。其中，每一个工作台对零件执行某一个操作，然后加工完的零件被传送到下一个工作台，它在前一工作台的位置被另一个未加工过的零件所取代。另一个例子是一个工艺流程。其中，各种各样大量的干燥材料被称重、组合、混合，最终被输出。

顺序控制能够以许多方式实现，包括机电继电器和各种各样的气体、流体及固态装置。然而，这一章集中讲解了主要用于实现开关控制的数字化计算机系统。这些专用的计算机被称为可编程逻辑控制。

鉴于前面的讨论，我们弄清楚了控制器的作用是连续地扫描其程序、根据开关的输入状态设定输出为开或关状态。因此，控制器中必须应装入能够定义所需切换顺序的程序。

课文 B　可编程逻辑控制器的连接

可编程逻辑控制器的型号不同，信号连接和编程标准也会略有变化，但是它们在 PLC 编程方面有很大的相似性，使我们可以在这里作"一般性的"介绍。图 10-5 显示了一个简单的可编程逻辑控制器（PLC）的正视图。两个标有 L1 和 L2 的螺丝接线端连接到 120 伏交流电上为 PLC 内部电路提供电源。左边的六个螺丝接线端连接到输入装置上，每个标有"X"的接线端都代表一个不同的输入"通道"。左下角的螺丝接线端是"公共端"，通常连接到 120 伏交流电源的 L2（中性线）上。

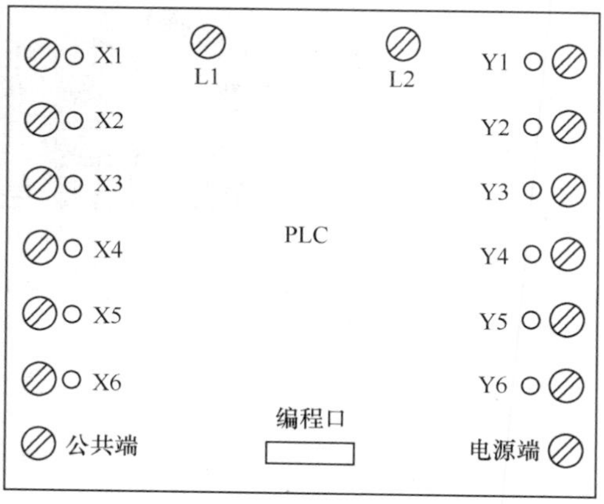

在 PLC 外壳内部，每个输入接线端和公共端之间有一个光电隔离装置（发光二极管）。当有 120 伏交流电分别加在输入端和公共端时，光电隔离装置的计算机电路（通过光敏晶体管接收发光二极管的光信号）提供"高电平"逻辑信号。PLC 的前部面板上的发光二极管用于指示"激活的"输入（见图 10-6）。

图 10-5

通过把"电源端"连到标有"Y"的输出端，PLC 的计算机电路激活开关装置（晶体管、双向可控硅，甚至是机电式继电器），生成输出信号。"电源端"通常连到 120 伏交流电电源的 L1 侧。和每个输入一样，PLC 的前部面板上有发光二极管用于指示"激活的"输出（见图 10-7）。

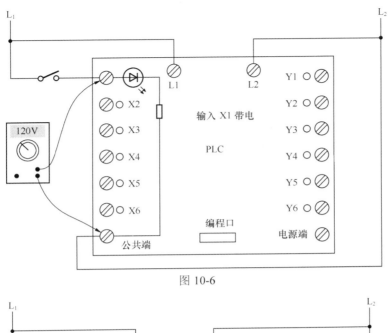

图 10-6

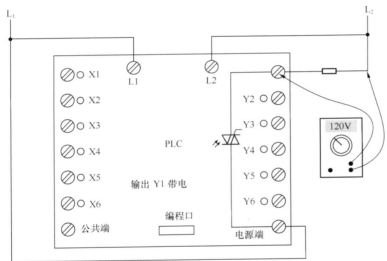

图 10-7

以这种方式，PLC 就能与实际的装置如开关和电磁阀连接。

控制系统的实际逻辑是通过计算机程序在 PLC 内建立起来的。这个程序给出在哪种输入状态下启动哪个输出命令。尽管程序本身显示为有开关和继电器符号的梯形图逻辑关系，但是并没有实际的开关触点或继电器线圈在 PLC 内工作，产生输入和输出的逻辑关系。如果你愿意，可以想象这些触点或开关的存在。通过连接到 PLC 编程口的个人计算机可以输入和查看程序。

看一下下面的电路和 PLC 程序（见图 10-8）。

当按钮开关没有激活（没按下）时，没有"电"送到 PLC 的输入端。根据下面的程序，常开触点与 Y1 线圈串联，没有"电"送到 Y1 线圈。因此，PLC 的 Y1 没有输出不带电，连接到 Y1 上的指示灯不亮。

然而，如果按下按钮，就有电送到 PLC 的 X1 输入端。程序中显示的任何一个 X1 触点都会处于激活（常闭）状态的话，就好像它们是因为继电器线圈"X1"通电而被激活的继电器。在这种情况下，带电的 X1 输入会使得常开触点 X1"闭合"，同时把"电"送到 Y1 线圈。当程序中的 Y1 线圈"带电"时，实际的 Y1 输出就会带电，连到输出端的显示灯亮（见图 10-9）。

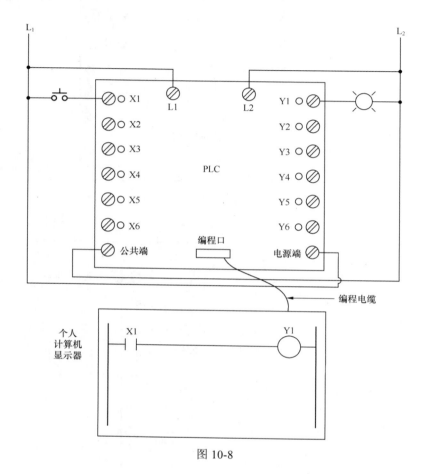

图 10-8

必须明白的是在个人计算机显示器上显示的 X1 触点、Y1 线圈、连接线以及"电"都是虚拟的。它们并不是以真实的电气元件存在的，而是在计算机程序中以指令形式（只是一个软件）存在的，就像真正的继电器动作框图。

此外，由于 PLC 中的每个输出信息只存在于存储器中，所以可通过输出状态（Y）来分配 PLC 程序中的触点。如下面的电机启动控制电路系统（见图 10-10）。

连到输入 X1 的按钮开关作为"启动"开关，而连到输入 X2 的按钮开关作为"停止"开关。程序中的另一个触点，被称为 Y1，将输出线圈状态作为自锁触点，这样电动机接触器在"启动"按钮释放后会继续带电。你可以看到常闭触点 X2 出现在带颜色的框中，表明处于闭合"带电"状态。

如果我们按下"启动"按钮，那么输入 X1 将会带电，从而"关闭"程序中的 X1 触点，给 Y1 线圈送电，使得 Y1 输出带电，同时把 120 伏交流电加到了真实的电机的接触器线圈上。并行的 Y1 触点也会"闭合"，这样"电路"会锁在带电状态（见图 10-11）。

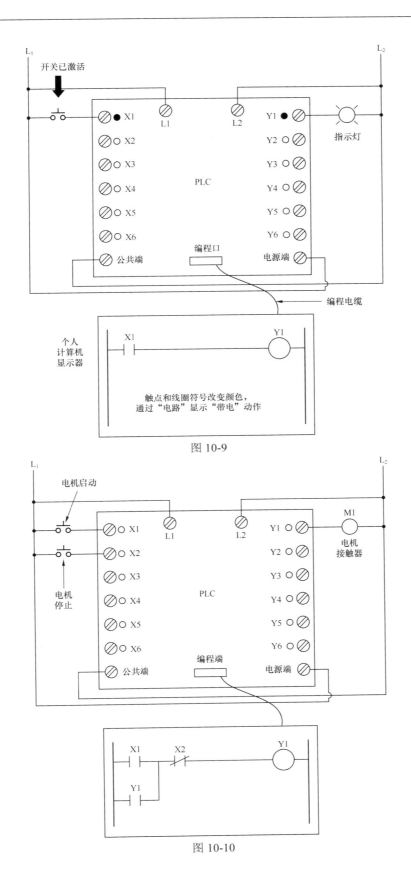

图 10-9

图 10-10

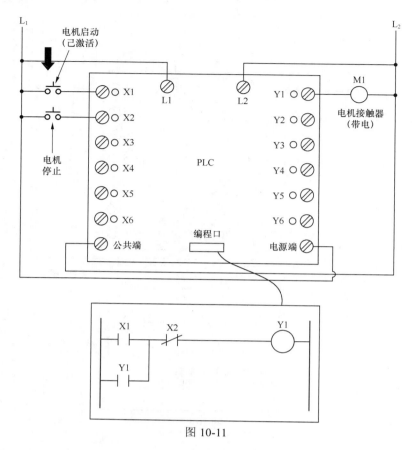

图 10-11

　　现在，如果我们释放"启动"按钮，那么常开触点 X1 就会返回到"打开"状态，但是电机会继续运行，因为自锁触点 Y1 继续给 Y1 线圈"连续性""供电"，这样 Y1 输出保持带电状态。

　　为了使电机停止运动，我们需立刻按下"停止"按钮，这样 X2 输入带电，同时常开触点"打开"，Y1 线圈停止供电。

　　当"停止"按钮被释放时，输入 X2 不带电，同时返回到正常的"闭合"状态。然而，电机在激活"启动"按钮前不会重新开始运行，因为自锁的 Y1 线圈已经失效。

参考文献

[1] 刘宇. 机电一体化专业英语. 天津：天津大学出版社, 2008.

[2] 刘晓莉, 苏雪, 邓青. 机电一体化及数控专业英语. 北京：人民邮电出版社, 2008.

[3] 杨成美. 机电专业英语. 北京：高等教育出版社, 2008.

[4] 侯继红. 机电专业英语. 北京：高等教育出版社 2008.

[5] 朱丽芬. 机电专业英语. 北京：中国电力出版社, 2008.

[6] 徐存善. 机电专业英语. 北京：人民邮电出版社, 2007.

[7] 鲍海龙. 机电专业英语. 北京：机械工业出版社, 2006.

[8] 朱德云. 机电专业英语. 北京：机械工业出版社, 2002.

[9] 徐存善. 机电专业英语. 北京：中国劳动社会保障出版社, 2007.

[10] 宋主民. 机电一体化专业英语. 北京：机械工业出版社, 2009.

[11] kudz, Myer. Mechanical Engineeis Handbook. 2nd Editiom. John Wiley & sons, 1998.

[12] 图片来源网页：http://vernier-caliper.com/.

[13] 图片来源网页：tresnainstrument.com.

[14] http://www.advic.co.th/teclock_products.html.

[15] 黄星. 图解英汉数控技术词典. 北京：化学工业出版社, 2012.

[16] 图片来源网页：http://www.virginia.edu/art/studio/safety/sculpture/mstools/benchgrinder.htm.